AMORPHOUS SILICON CARBIDE THIN FILMS: DEPOSITION, CHARACTERIZATION, ETCHING AND PIEZORESISTIVE SENSORS APPLICATIONS

MATERIALS SCIENCE
AND TECHNOLOGIES

Additional books in this series can be found on Nova's website
under the Series tab.

Additional E-books in this series can be found on Nova's website
under the E-book tab.

MATERIALS SCIENCE AND TECHNOLOGIES

AMORPHOUS SILICON CARBIDE THIN FILMS: DEPOSITION, CHARACTERIZATION, ETCHING AND PIEZORESISTIVE SENSORS APPLICATIONS

MARIANA AMORIM FRAGA

Nova Science Publishers, Inc.
New York

LIBRARY OF CONGRESS CATALOGING-IN-PUBLICATION DATA

Fraga, Mariana Amorim.
  Amorphous silicon carbide thin films : deposition, characterization, etching, and piezoresistive sensors applications / Mariana Amorim Fraga.
    p. cm.
  Includes bibliographical references.
  ISBN 978-1-61324-774-7 (softcover)
  1. Silicon carbide. 2. Silicon-carbide thin films. 3. Amorphous semiconductors. I. Title.
  TK7871.15.S56F72 2011
  621.3815'2--dc23
                          2011017290

Published by Nova Science Publishers, Inc. † New York

DEDICATION

To my mother, *Antonia Oliveira Amorim Fraga (in memoriam)*

CONTENTS

PREFACE

Silicon carbide (SiC) has been described as a suitable semiconductor material to use in MEMS and electronic devices for harsh environments. In recent years, many developments in SiC technology as bulk growth, materials processing, electronic devices and sensors have been shown. Moreover, some studies have also shown the synthesis, characterization and processing of crystalline SiC films. However, few works have investigated the potential of amorphous silicon carbide (a-SiC) thin films for sensors applications.

The goal of this book is to present fundamentals of amorphous silicon carbide thin films and their applications in piezoresistive sensors for high temperature applications. The book is introduced with a chapter on the motivations for using amorphous SiC films in microelectronic devices and MEMS sensors. Chapter 2 presents a literature review on methods of obtaining SiC in bulk and thin-film. The PECVD and RF magnetron sputtering processes are emphasized. Some techniques used to investigate chemical, structural, morphological, mechanical and electrical properties of thin films are presented. In this chapter, the SiC processing steps are also briefly described. A description of experimental procedures for deposition, characterization and etching of SiC films is presented in Chapter 3. The analyses of the results are discussed in Chapter 4. In Chapter 5 is shown an overview on SiC piezoresistive sensors. A methodology to characterize piezoresistive properties of thin films is described. The steps of design, fabrication and characterization of a prototype of piezoresistive pressure sensor based on SiC film are also shown. Finally, the conclusions of this book are presented in Chapter 6.

ACKNOWLEDGMENTS

First, I would like to thank my father, Orlando Bastos Fraga, for his support and incentive. Then, I would like to thank my boyfriend Leandro Koberstein for patience and love throughout the period of the writing of this book. He was always there for me.

I have to thank the collaboration of the researchers Marcos Massi, Rodrigo Pessoa, Humber Furlan, Ivo Oliveira, Homero Maciel and Luiz Rasia for the development of this book. In addition, I would like to thank all my colleagues from Plasma and Processes Laboratory of the Technological Institute of Aeronautics (LPP-ITA), the researchers Angelo Gobbi and Maria Helena Piazzetta from Brazilian National Synchrotron Light Laboratory (LNLS), Fernando Josepetti Fonseca, and Sebastião Gomes dos Santos from University of São Paulo (USP).

I would like to thank also some relatives and friends: Célia Fraga, Aristides Amorim, Rosalvo Magalhães, Maria do Carmo Magalhães, Maysa Magalhães, Ádma Magalhães, Márcia Magalhães, Roberto Magalhães, Rogério Magalhães, Fernando Lopes, Cláudio Dias, Petrina Dias, Neusa Koberstein, Franciso Léo, Teresa Léo, Roseli Léo, Sidney Fernandes, Carlos Eduardo Bicalho, Rosane Bicalho, Shirley Wakavaiachi, and Janderson Rodrigues.

Finally, I would like to acknowledge the financial support of the Brazilian research agencies CNPq, CAPES, FINEP and FAPESP; and the Institutions that have provided their infrastructure for the experiments: Technological Institute of Aeronautics (ITA), Microfabrication Laboratory (LMF-LNLS), Institute for Advanced Studies (IEAv), Center of Semiconductor Components (CCS-UNICAMP), Faculty of Technology of São Paulo (FATEC-SP), Laboratory of Technological Plasmas (LAPTEC-UNESP), Associate

Laboratory of Sensors and Materials (LAS-INPE), Laboratory of Material Analysis by Ionic Beams (LAMFI-USP), Laboratory of Optical Properties (LAPO-UFBA), and Laboratory of Integrated Systems (LSI-USP).

Chapter 1

INTRODUCTION

Aeronautical and aerospace systems have benefited from advances in microelectronics that have made possible the fabrication of devices with size, mass and reduced power. However, these systems usually operate in harsh environmental conditions such as high temperatures and intense radiation, which limits the use of some semiconductor materials. It is known, for example, that silicon devices have satisfactory performance at temperatures lower than 150°C and the SOI (Silicon-On-Insulator) at temperatures below 500°C [1]. These limitations have become a challenge to consolidate the process of miniaturization because besides of small the devices should be able to operate well in harsh environments. This sparked interest in the study of new semiconductor materials such as silicon carbide (SiC), which have excellent thermal and chemical stability as well as compatibility with microfabrication technology based on silicon [2].

The study of SiC for electronic applications has intensified since 1959 when Shockley, inventor of the bipolar transistor, recognized the SiC as a suitable material for manufacturing microelectronic devices capable of withstanding high temperatures and intense radiation [3]. Since then, several techniques to grown SiC bulk and films have been developed. This interest is justified by the excellent properties of the SiC, as for example high chemical stability, high tensile strength, high mechanical hardness and wide band-gap [4]. Because of these properties, SiC has been used as base material in the fabrication of different MEMS (Micro-Electro-Mechanical Systems) and MOEMS (Micro-Opto-Electro-Mechanical Systems) devices for different automotive, petrochemical, aeronautical and aerospace applications [5].

Among the different types of MEMS devices, one of the most studied are the piezoresistive sensors.

The piezoresistive effect is the change of resistivity in a semiconductor caused by applying mechanical stress [6]. The magnitude of this effect is directly related to temperature, so there is growing interest in to characterize the piezoresistive properties of materials with good thermal stability. The literature shows that the SiC due to its excellent thermal and chemical stability is a good candidate for applications at temperatures up to 600°C [1,7].

An important parameter to characterize the piezoresistive properties of a material is study the gauge factor (GF), which relates the variation of electrical resistance with deformation of the material. When compared to monocrystalline silicon, the SiC has a low GF. This means that SiC only responds to large strain. Moreover, while the GF of silicon decreases significantly with temperature increases, the SiC undergoes minor variations.

One difficulty for the development of sensors based on SiC is the characterization of the piezoresistive effect in this material. The literature shows that the piezoresistive effect in semiconductors is highly anisotropic having a dependency direct on types and concentrations of dopants and on the crystallographic orientation of the material obtained [8]. It is known that the SiC has about 200 different polytypes that can present different gauge factors. Another factor that limits the use of SiC as a material for sensors is a difficult bulk micromachining technology [9]. To overcome this problem some processes have been developed as the deposition of SiC films on Si or SOI (Silicon-On-Insulator) substrates [10]. The advantage of use of these substrates is that the wet and dry anisotropic etching of Si is well known in the literature.

This book presents studies on un-doped and nitrogen-doped amorphous silicon carbide (a-SiC) thin films deposited on Si substrates. The main steps required performing the deposition, characterization and ecthing of a-SiC films are described. Moreover, a methodology to develop piezoresistive sensors based on these films is also presented.

Chapter 2

AN OVERVIEW ON SILICON CARBIDE (SIC)

Silicon carbide (SiC) is a semiconductor material with great potential for use in devices for high temperature and high power, which can be obtained in crystalline or amorphous form. SiC presents a special case of polymorphism (different types of crystalline structures) called polytypism. In this case, the change in stacking sequence of atoms acts only in one specific direction in the three-dimensional network. This property allows an almost infinite variation in the arrangements of atoms, which result in more than 200 polytypes of SiC [11]. However, few polytypes have good semiconducting properties. Among them, stand out the 4H-SiC, 6H-SiC, 2H-SiC and 3C-SiC [12]. Figure 2.1 shows the stacking of layers of these SiC polytypes. The letters H and C refer to the symmetry of the crystal, H for hexagonal and C for cubic (Figure 2.2) and the numbers refer to the SiC layers are stacked differently to form a unit cell.

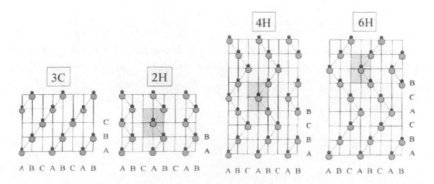

Figure 2.1. Stacking order of the main polytypes of SiC.

Although SiC polytypes have the same chemical composition, some properties have different values. For example, the band-gap energy of 3C-SiC is 2.36 eV while the 4H-SiC is of 3.23 eV. Table 2.1 compares the properties of the main polytypes of SiC with Si, Ge and diamond.

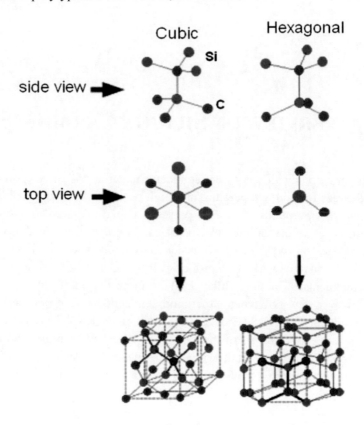

Figure 2.2. Main crystalline structures of the SiC.

As can be observed, the SiC polytypes have band-gap energy and elastic modulus considerably higher than the of Si and lower than the of diamond. This indicates that the SiC has intermediate properties between these materials.

Table 1.1. Comparison among the properties of SiC polytypes, Si, Ge and diamond

Properties	3C-SiC	4H-SiC	6H-SiC	Si	Ge	Diamond
Bandgap (eV)	2.36	3.23	3.0	1.11	0.67	5.5
Breakdown voltage (MV/cm)	1.0	3.0 to 5.0	3.0 to 5.0	0.3	0.1	10
Thermal conductivity (W/cm-K)	3.6	3.7	4.9	1.3	0.58	20
Elastic or Young´s modulus (GPa)	448	448	448	190	80	1035
Dielectric constant	9.7	9.7	9.7	11.7	15.8	5.5

2.1. TECHNIQUES USED FOR OBTAINING SiC

2.1.1. Bulk Growth

Acheson developed the first process of obtaining SiC bulk in 1891. In this process, the SiC is produced by the reaction of silica with carbon at about 2700°C. In 1955, Lely has developed a method of production based on the sublimation of SiC in argon atmosphere at temperatures between 2200 and 2700°C. The Lely process was the first to be used for production of SiC semiconducting.

In 1977, a group of researchers at the University of Saint-Petersburg modified the Lely process making possible the attainment of SiC at 2200°C at a pressure below atmospheric pressure. This process is known as modified Lely and is currently the method most used by companies that manufacture SiC bulks. In Figure 2.3 is summarized the history of SiC technology for bulk. The SiC wafers of the polytypes 4H-SiC and 6H-SiC become commercially available only in 1991.

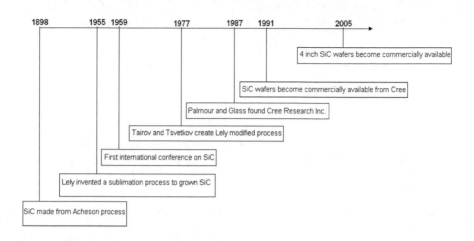

Figure 2.3. Brief history of SiC technology for bulk.

In 1998, Neudeck from NASA presented a study on SiC technology [13]. One of the main conclusions of this study is that compared to silicon and GaAs wafers the commercial SiC wafers were small, expensive, and generally of inferior quality. Besides, the SiC wafers exhibited high densities of crystalline defects and significantly rougher surfaces [13].

In 2004, Cree Research Inc. presented studies showing that in the period among 1997 to 2004 the density of defects on the 4H-SiC wafers was significantly reduced, while the wafer diameter doubled [14,15]. These studies also showed that the properties of the 4H-SiC are higher than those of Si and GaAs. However, despite the reduction of defects, increasing the diameter and the excellent properties of 4H-SiC, this wafer is still underutilized because of the high cost.

In relation to the 3C-SiC polytype, in 2000, the Japanese company Hoya was the first to develop 3C-SiC wafers using a low temperature process (1000°C) to grown 3C-SiC on 150 mm Si wafers. When the thickness of the film achieves 300 μm, the Si substrate is removed [16].

Even with the recent advances in SiC technology to grown bulk, the high cost of the SiC wafers still have motivated several studies on processes to produce SiC films on Si or SOI wafers. The next section presents some of these processes.

2.1.2. Thin Film Growth

The use of SiC thin films in semiconductor devices has several advantages such as easy integration with integrated circuit technology, low cost and possibility of fabrication of structures at the microscopic level [17]. Although the semiconducting properties of SiC are known since the 50s, initially SiC thin films were produced only for use in coating of cutting tools to minimize the wear caused by operating conditions more severe [18].

In 1968, the first papers on the electrical and optical characterization of SiC films obtained by PECVD [19], CVD [20] and sputtering [21] were published. The results shown in these papers have motivated several studies on the application of SiC films in semiconductor devices.

Several techniques can be employed for the deposition of SiC films on different substrates. The Si and SOI (Silicon-On-Insulator) substrates are most often used because the microfabrication technology in these substrates is already consolidated. Crystalline SiC films are obtained by techniques that use temperatures higher than 1000°C as the CVD (Chemical Vapor Deposition), MBE (Molecular Beam Epitaxy) and ECR (Electron Cyclotron Resonance) [22]. The main disadvantages of these techniques are the high temperatures involved in the deposition process which makes it difficult the use of these films in conjunction with conventional processes for microelectronics. An alternative to this problem is to use plasma assisted processes as PECVD (Plasma Enhanced Chemical Vapor Deposition) and sputtering, which made possible to obtain SiC films at low temperatures (below 400 ° C). However, the low temperatures used in these processes often results in obtaining amorphous SiC films that have distinct properties of the crystalline. In this context, many studies have been performed on the deposition and characterization of amorphous films aiming their application in microelectronic devices [23,24]. In this book are presented the a-SiC films obtained by PECVD and RF magnetron sputtering.

2.1.2.1. Plasma Enhanced Chemical Vapor Deposition (PECVD)

The PECVD is a thin film deposition technique based on cold plasma. This type of plasma is characterized by a lack of thermodynamic equilibrium between the electron temperature and the ions temperature, i.e.; the kinetic energy (which is proportional to temperature) of electrons is much greater than that of ions [25]. In this process, gas precursor molecules or reactants are broken by an applied electric field between two electrodes and the radicals resulting from this decomposition react chemically to form a solid film. Figure 2.4 illustrates the deposition mechanism of this technique [26].

In the deposition of SiC films by the PECVD, the precursor gases most used are SiH_4 as a silicon source and CH_4 as carbon source. A PECVD deposition system is schematically represented in Figure 2.5. The cathode, used as substrate holder, is connected to a source of radio frequency through an impedance matching. The walls of the chamber form the anode (grounded electrode). The plasma is generated by collisions of electrons, accelerated by the RF field, with the atoms and/or molecules from the precursor atmosphere causing further ionizations through several collisions.

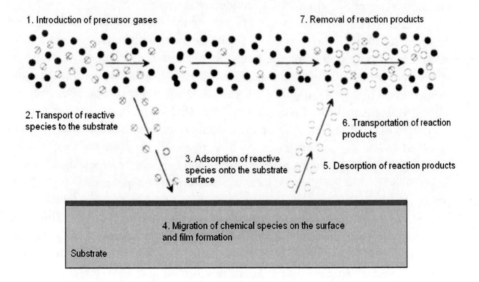

Figure 2.4. Scheme of the different processes involved in the deposition mechanism.

The process of bombarding of the substrate is caused by the acceleration of ions through the cathode sheath. The main parameters that control the energy of bombardment of the substrate surface are the electric field of the sheath and the pressure inside the reactor.

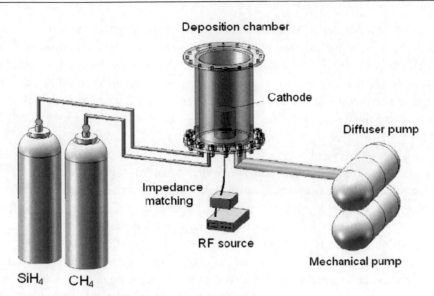

Figure 2.5. Schematic representation of a PECVD system.

2.1.2.2. Magnetron Sputtering

The sputtering technique can be defined as a process that occurs when a particle, with high kinetic energy, reaches a surface causing the ejection of atoms or molecules [27]. It is necessary to place the substrate in the direction of the target to occur the deposition of the atoms ejected from the target onto substrate surface.

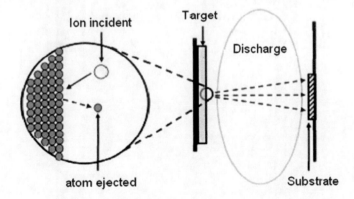

Figure 2.6. Schematic representation of the sputtering process.

The sputter deposition mechanism is schematically represented in Figure 2.6. To ensure the efficiency of the sputtering process is essential that the incident particles in the target have atomic dimensions, so for a very small particle like an electron, for example, does not guarantee that the momentum transfer to the atoms of the material that constitute the target is sufficient to occur thinning, and, consequently, the deposition of the film on the substrate [28].

There are several ways of performing a sputtering process. The simplest model called conventional sputtering is performed in a vacuum chamber with electrodes connected to a source of high voltage direct current. The target is placed on the cathode and the substrate is positioned a certain distance from the cathode. Figure 2.7 (a) illustrates this process of conventional sputtering deposition [29]. It is observed that the ionization reactions occur throughout the plasma, which is spread across the chamber so that many ions formed do not reach the target surface resulting in a low deposition rate. One way to solve this problem is to use a magnetic field placed near the cathode. In this case the technique is called magnetron sputtering. The magnetron confines the plasma near the cathode surface as shown in Figure 2.7 (b).

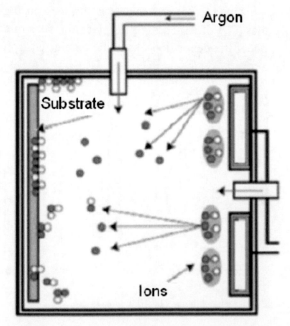

a

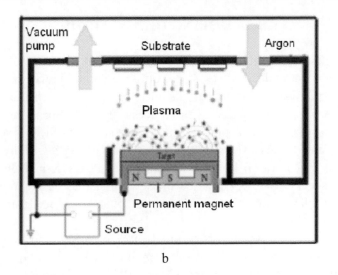

b

Figure 2.7. Deposition by sputtering: (a) conventional and (b) magnetron.

In this type of sputtering process, a DC source is used to deposit metals and an RF source to deposit thin films insulating or semi-insulating (process known as RF magnetron sputtering) [30]. Figure 2.8 shows a schematic illustration of an RF magnetron sputtering reactor to deposit SiC films using a SiC target in argon atmosphere.

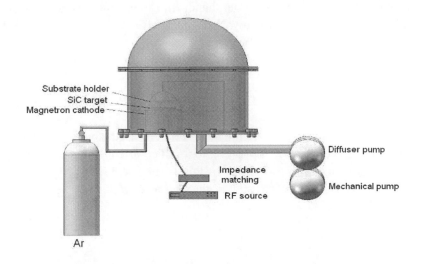

Figure 2.8. Schematic representation of a RF magnetron sputtering system.

2.2. DOPING OF SiC FILMS

Until the 1970s, the main difficulty for the fabrication of devices in amorphous semiconductors was to perform the doping process efficiently. It was believed that the structural disorder of these materials would make impossible the precise control of the process. In 1975, Spear and LeComber published a paper on the doping of amorphous silicon (a-Si) [31] and in 1977 Anderson and Spear investigated the electrical properties of hydrogenated amorphous silicon (a-Si: H) alloys obtained by PECVD, including a-SiC:H [32]. The sucess of these works led the several studies on the application of amorphous silicon alloys in optical sensors, photovoltaic cells and thin film transistors (TFTs).

In the 1980s, several papers on doping of a-SiC and a-SiC:H were published [33]. In these works, in situ doping of SiC films, n-type and p-type, were carried out respectively by the addition PH_3 (or N_2) and B_2H_6 the mixture of precursor gases during deposition process. However, the doping process of SiC was consolidated only in the 1990s, with the use of post-deposition techniques as ion implantation and thermal diffusion.

Currently, doping of amorphous and crystalline SiC films is mainly carried out by ion implantation. The advantage of this technique is that it can be performed in selective regions besides be a well controlled process. The technique of thermal diffusion of aluminum is also widely used for doping of amorphous SiC films. On the other hand, it should not be used to dope crystalline SiC films due to the low diffusion coefficient of impurities of this material [34].

2.3. TECHNIQUES USED TO CHARACTERIZE SiC FILMS

Several characterization techniques can be used to investigate the composition, chemical bonds, crystallographic orientation, morphology, resistivity and hardness of SiC films. In this section will be briefly described only the techniques that were used in the characterization of the SiC films studied in this book.

2.3.1. Rutherford **Backscattering Spectrometry** *(RBS)*

The RBS technique allows analyzing the chemical composition of a material. This technique consists of measuring the energy of particles of a monoenergetic beam after collisions with atoms of the target (sample) that will be examined and which were backscattered [35]. During the process, the incident ions lose energy by collisions with atoms of the sample being that the reduction rate of energy of the backscattered particle depends on ratio of the incident particle mass and of the target atom. This allows identifying what the atoms mass of the target and therefore what is the corresponding chemical element.

Besides the chemical composition, the RBS technique allows also determine the thickness of the sample. When compared to other techniques for chemical composition analysis, RBS is not as sensitive to thin layers and low concentrations. However, the RBS technique has other advantages as does not to destroy the sample, rapid analysis (it is not necessary a calibration standard) and a simple interpretation of the spectra [35].

Initially, the RBS spectra show only the elements that constitute the sample without determine the quantity of each one. In order to obtain quantitative information of film composition is necessary to perform a simulation on the spectrum extracted from the experimental analysis. The goal of the simulation is that the simulated spectra overlap the experimental. From the overlap of spectra is possible to determine quantitatively the elements in the sample. There are some softwares for simulation of RBS spectra being that the most used are the RUMP [36] and the SIMRA [37].

RBS measurements of a-SiC films that will be presented in Chapter 4 of this book were carried out using a 2MeV He^+ beam. The RBS spectra showed were analyzed using a RUMP code (RBS spectroscopy analysis package) developed by Cornell University [36].

2.3.2. Fourier Transform Infrared Spectroscopy (FTIR)

The FTIR technique studies the interaction of infrared radiation with the matter. The emitted infrared radiation covers the electromagnetic spectrum in three energy ranges: the near infrared (13000 to 4000 cm^{-1}), mid infrared (4000 to 200 cm^{-1}) and far infrared (200-10 cm^{-1}). The energy range commonly used in characterization of materials is the mid infrared [38].

The FTIR technique is based on the following principle: when the frequency of the incident infrared radiation equals molecule's vibration frequency that constitute the material, the radiation can be absorbed causing the vibration of the chemical bonds. The infrared spectrum consists of absorption bands corresponding to frequencies of bonds of the atoms that compose the material. The IR spectrum is a two-dimensional representation of the characteristics of absorption of a molecule that appears as bands or peaks and can be described in terms of three variables: position, intensity, and shape [38]. The intensity is defined by the amount of energy absorbed by the sample, when exposed to infrared radiation, which is proportional to the number of molecules that interact with the radiation observed. The intensity is given in terms of transmittance (T), which measures the ratio between the radiant power transmitted (I) and the radiant power incident (I_0).

The main advantage of FTIR technique is to use a Michelson interferometer. This interferometer enables that the infrared spectroscopy has greater sensitivity and speed in making the measure, greater intensity in the detector, automatic calibration, no stray light and no heating. The interferogram is formed by the sum of all waves of different frequencies and amplitudes that provides the information about the sample under analysis, however, so as not resolved and must be transformed into a diagram, spectrum, which relates the intensities with the corresponding frequencies [39]. The mathematical treatment to achieve this goal is the Fourier transform.

Infrared spectroscopy of a-SiC films produced was carried out by a Perkin Elmer spectrum 2000 Fourier transform infrared spectrometer in the range 400–4000cm^{-1}. A (100) silicon wafer was used as reference. The results obtained will be shown in Chapter 4.

2.3.3. Raman Spectroscopy

Radiation can interact with matter by processes of absorption or scattering being that the latter can be elastic or inelastic. In the process of elastic scattering, or Rayleigh scattering, the incident photons are elastically scattered, i.e. the incident photons have the same length as the photons absorbed. In inelastic scattering or Raman scattering, first reported in 1928 by the Indian physicist Chandrasekhara Raman Vankata, the energies of the incident and scattered photons are different [39].

The Raman scattering can be understood as follows: a molecule that is in some state of vibration, not necessarily the key, and absorbs a photon of energy $h\upsilon_i$ that turns into an intermediate state (or virtual), immediately makes a transition to a state of energy higher than the initial state emitting (scattering) photon energy $h\upsilon_f$, so that $h\upsilon_f < h\upsilon_i$ (Stokes line). In order to conserve energy, the difference $h\upsilon_i - h\upsilon_f = h\upsilon_{cb}$ excites the molecule to a vibrational energy level higher. If the molecule is initially in an excited vibrational state (which can happen, for example, if the sample is heated), then to absorb and emit a photon, it can decay into a state of lower energy. In this case $h\upsilon_f > h\upsilon_i$ means that some vibrational energy of the molecule was converted into photon energy spread, so that $h\upsilon_f - h\upsilon_i = h\upsilon_{ba}$ (anti-Stokes line). The Stokes and anti-Stokes waves are a direct measure of the vibrational energy of molecules. In both cases the energy difference between the incident and scattered photons is called the Raman shift and corresponds to different energy levels of the specific sample studied.

The Raman spectrum lists the number of scattered photons as a function of Raman shift around a wavelength of incident laser radiation. A Raman spectrum is obtained when a monochromatic light from a laser incides on the sample to analyze. The scattered light is dispersed by a diffraction grating in the spectrometer and its components are collected in a detector that converts light intensity into electrical signals that are interpreted using a computer [40].

Raman spectroscopy of the a-SiC films produced was performed with a Renishaw 2000 system using an Ar^+-ion laser ($\lambda = 514$ nm) with power 0.6 mW. The Raman spectra were obtained at room temperature in the range of 400 to 2000 cm^{-1}. The results are shown in Chapter 4.

2.3.4. X-Ray Diffraction (XRD)

The x-ray diffraction is one of the main techniques for microstructural characterization of crystalline materials by dispenses complex methods of sample preparation, to be nondestructive and to allow the analysis of materials consisting of a mixture of phases [41]. In this technique, beams of monochromatic x-rays are used to determine the interplanar spacing of the material analyzed.

The XRD technique uses the following principle: the x-rays that achieve a material can be scattered elastically without energy loss by electrons of an atom (or coherent scattering dispersion). After the collision with an electron, the photon x-ray changes its trajectory while maintaining the same phase and incident photon energy with this the electromagnetic wave is instantly absorbed by the electron and then reemitted hence each electron acts as a center of x-ray emission [42]. If the atoms that generate this scattering are arranged in a systematic manner as in a crystalline structure and the distances between them are close to the wavelength of incident radiation, it is verified that the phase relationships between the scatterings become regular and effects of x-ray diffraction can be observed at various angles. The conditions for the occurrence of x-ray diffraction are defined by Bragg's law:

$$n\lambda = 2d.sen\theta \qquad\qquad (2.1)$$

Where n is an integer number (diffraction order), λ corresponds to the wavelength of incident radiation, d is the inter-planar distance for the set of planes hkl (Miller index) and θ is the angle of diffraction x-ray beam measured between the incident beam and crystalline plans.

In the spectra obtained by X-ray diffraction the position, intensity and width of the peaks provide the information about the structure of the material analyzed.

XRD spectra of the a-SiC films that will be presented in Chapter 4 of this book were obtained using diffractometer Philips PW 1380/80 in the θ–2θ scan configuration using Cu-Kα radiation.

2.3.5. Atomic Force Microscopy (AFM)

The atomic force microscopy (AFM) is a technique that allows obtaining real images, in three dimensions, of the surface topography with a spatial resolution approaching the atomic dimensions [43]. The working principle of an atomic force microscope can be summarized as follows: the sample is placed on a piezoelectric ceramic (scanner) that serves to position it, a tip supported by a cantilever traverses the surface of the sample, on the cantilever incides the light from a laser that is reflected, falls into a mirror and then reaches a detector. The function of the detector is to measure the deflections of the cantilever caused by the roughness of the sample when the tip scans it. The results of these measurements are transferred to a computer using a specific program converts the information into an image of the surface.

The main component of a microscope AFM is the tip (or probe) that scans the sample surface under study. The forces involved in the scanning of the surface are of two types: those of attraction (Van der Waals forces) that act at distances less than 100 nm and repulsion forces that come from the Pauli exclusion principle and act when the tip comes very close to leaving the surface atoms of the probe and sample are close enough that begin to repel [43].

The modes to obtain AFM images can be classified as: contact, intermittent contact and non-contact, depending on the forces between tip and sample. In AFM non-contact mode, the cantilever bends towards the sample (attractive region) while in contact mode is the opposite: the cantilever bends away from the sample (repulsive region). The intermittent contact occurs when the cantilever is forced to oscillate at a certain frequency and at a certain height of the sample, so the cantilever touches the sample only periodically. In this mode, the force acting is sometimes attractive and sometimes repulsive.

Some external factors such as humidity, temperature, and vibrations can affect the analysis of a sample by AFM.

AFM images of the a-SiC films were obtained by SPM-9500J3 system in dynamic mode. The AFM results are presented in Chapter 4.

2.3.6. Four Points Probe

The method of four points is the most commonly used to determine the electrical resistivity of thin films or bulks of conductive and semiconductive materials. The main advantage of this method is the simplicity of the measure, and the fact it is not necessary a good ohmic contact between the electrode and the sample [44].

The experimental apparatus of this technique consists in four vertical electrodes, whose tips are in the same plane, co-linear oriented, and regularly spaced. The two outer tips are used to transport current (I) and the two ends inside to measure the voltage (V).

As the tips are equally spaced the resistivity can be calculated by:

$$\rho = 2\pi S \frac{V}{I} \tag{2.2}$$

Where S is the spacing between the tips, V is the measured voltage and I is the current injected. This equation is only valid for a semi-infinite substrate [45].

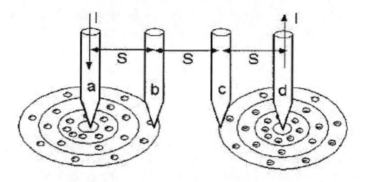

Figure 2.9. Flow of charged particles in a thin film.

To determine resistivity of thin films is necessary to add a geometric correction factor and the following considerations should be made: the thickness (t) of the thin film should be much smaller than the spacing between the points (S) while the sample size that will be measured should be much larger than this spacing. It is known that the current propagation on the film is in rings shape (see Figure 2.9) with perimeter of $2\pi r$ and thickness equal the film thickness.

As can be observed in Figure 2.9, the current (I) is injected into the system through the tip (a) and it flows up to tip (d). The potential difference is measured between the tips (b) and (c). Thus, integrating the differential resistance on the length between these two electrodes and considering that either external tip induces the voltage between them, the resistivity for thin films is given by [45]:

$$\rho = \frac{\pi}{\ln 2} t \left(\frac{V}{I} \right) \tag{2.3}$$

Where $\pi/\ln 2$ is the geometric correction factor necessary to determine the resistivity of a thin film.

2.3.7. Nanoindentation

The technique of nanoindentation has a main objective of determining the hardness (H) and the elastic modulus (E) of solid surfaces or thin films. This technique consists carried out charge/discharge cycles as a function of penetration depth. The data produced are arranged in a load-displacement diagram, which is called charge-discharge curve. This curve describes the elastic-plastic behavior of the material. Some methods were developed to obtain the hardness and elastic modulus of a material [from] the charge-discharge curves. Currently, the Oliver and Pharr method is the most used [46].

The nanoindentation technique is widely used approach for the determination of hardness and elastic modulus of thin films, which can be obtained through an interpretation of experimentally obtained load–displacement curves during the indentation load /unload cycle.

In geral, nanoindentation measurements are performed using a Berkovich diamond indenter (a three-sided pyramid). The physical principles used to determine E and H are based on the Oliver-Pharr theory [46] where the fundamental relations to calculate hardness and reduced elastic modulus are:

$$H = \frac{P_{max}}{A} \qquad (2.4)$$

$$E_r = \frac{\sqrt{\pi}}{2} \frac{S}{\sqrt{A}} \qquad (2.5)$$

Where P_{max} is the peak load, S is the parameter known as the elastic contact stiffness and A is the projected contact area at that load that for a Berkovich tip is defined by:

$$A = 24.5 h_c^{\,2} \qquad (2.6)$$

Where h_c is the final plastic contact depth.

The elastic modulus of the material, E, is then calculated from E_r by the expression:

$$\frac{1}{E_r} = \frac{1 - v^2}{E} + \frac{1 - v_i^{\,2}}{E_i} \qquad (2.7)$$

Where v_i, v, E_i and E are the Poisson's ratio and Elastic modulus of the indenter and sample, respectively.

Elastic modulus and hardness of the a-SiC films produced were measured using a Hysitron triboindenter with a standard Berkovich diamond. The nanoindentation experiments were conducted under load control with load varying from 100 to 1800 µN and indentation depth did not exceed 10% of the film thickness. The results obtained are shown in Chapter 4.

2.4. SILICON CARBIDE PROCESSING

The semiconductor device fabrication, layer by layer, depends fundamentally on the definition of areas on the material that will be subjected to physical-chemical treatments that will result in the electronic components and interconnections. The lithography process is responsible for sizing, positioning and alignment of these successive layers. The other processing steps that are considered essential to the development of a device are oxidation, metallization, and etching. The next sections present brief descriptions of these steps in the manufacturing process of devices based on SiC.

2.4.1. Oxidation

Thermal oxides can be grown on SiC using the same methods used to oxide silicon substrates. However the mechanisms are different [47,48]. Silicon is achieved through direct oxidation of the substrates in oxidative environments. For SiC, the process is the same but the mechanism of oxide formation is more complex. The reaction between SiC and O_2 is as follows:

$$2SiC_{(s)} + 3O_{2(g)} \rightarrow 2SiO_{2(s)} + 2CO_{(g)} \qquad (2.8)$$

The oxidation rate of SiC is much lower than of the Si. For instance, a process used to grow a layer of 1.5 µm silicon oxide will grow only about 90 nm on 3C-SiC. Besides the low deposition rate, the interface SiO_2/SiC has a density of electrically active defects considerably high.

2.4.2. Metallization

The fabrication of the electrical interconnections is a critical step in the development of electronic and opto-electronics devices, especially in the small size devices with little available area for contact or high-density integration. When the goal is to develop a device for use in high temperatures, generally ohmic and Schottky contacts are made of metals with high [melting] temperatures such as Ni, Ti and W. In SiC devices, these metals are not widely used [49]. In these devices, the ohmic contact most used is the Al because is easily deposited by sputtering or evaporation. The main problem is that the Al melts at 600°C, which limits its application at high temperatures.

The best Schottky contact for SiC is the Au. As well as the Al, Au is easily deposited by sputtering. Although the Au has a high melting point it is not suitable for application at high temperatures due to the effect of electromigration, which is characterized by the gradual migration of metal atoms. This effect is greater the higher the current density and operating temperature [50]. Some metals as Ti enhance the resistance to electromigration of Au. So, in the device fabrication processes often Au layers are deposited on Ti layers. This procedure besides ensure good adhesion of Au, forms also a barrier to electromigration.

2.4.3. Etching

The chemical inertness of the SiC, which is considered one of its most attractive features for application in harsh environments, makes difficult the micromachining process. It is known that the crystalline and amorphous SiC are resistant to wet etching in solutions as KOH and HF [51]. This difficulty to etch the SiC in chemical solutions has stimulated the growing interest by use of techniques of plasma etching such as RIE (Reactive Ion Etching), to define the geometry of devices based on SiC.

In plasma etching of SiC, the gases more used are the fluorinated as SF_6, CHF_3 and CF_4. The reaction mechanism of SiC with fluorine is the following [52]:

$$Si + xF \rightarrow SiF_x \quad x \leq 1-4 \tag{2.9}$$

$$C + xF \rightarrow CF_x \quad x \leq 1-3 \tag{2.10}$$

As fluorinated gases are electronegative, during the etching process occurs a great capture of low energy electrons. This reduces the dissociation of the fluoride, which causes a reduction in the etching rate. To increase the dissociation of fluorine is added another gas to process, usually the oxygen [53,54,55]. This procedure increases the etching rate of the SiC. Other parameters that also influence the etching rate are the pressure and the RF power.

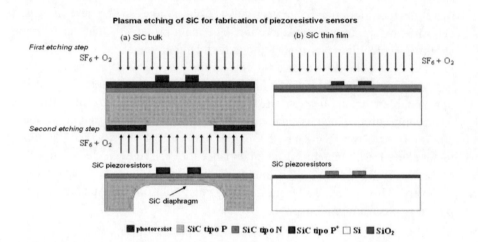

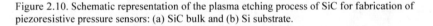

Figure 2.10. Schematic representation of the plasma etching process of SiC for fabrication of piezoresistive pressure sensors: (a) SiC bulk and (b) Si substrate.

In the fabrication of SiC piezoresistive sensors, the plasma etching process is a key step to produce the structures that constitute the sensor. When using SiC bulk, are two stages of plasma etching: one to form the piezoresistors and the other for the diaphragm (Figure 2.10 (a)). For the SiC film deposited on Si or SOI substrate the plasma etching is used only to fabricate the SiC piezoresistors (Figure 2.10 (b)). In this case, the diaphragm is usually formed by anisotropic etching of Si in KOH solution.

The main advantage of using Si substrate for fabrication of piezoresistive sensors is that the micromachining process in aqueous solutions such as KOH or TMAH is already established in the literature. The wet etching process, besides being simple and inexpensive, also enables precise control of the thickness of the diaphragm. When the SiC film is deposited on Si or SOI substrate, the SiC film (^{Figure} 2.11 (a)) or SiO_2 layer of the SOI substrate (Figure 2.11(b)) acts as a etch stop layer [56]. This is because SiC and SiO_2 have lower etching rate than silicon in alkaline solutions commonly used in the wet etching process. When the SiC film is deposited on an oxidized Si substrate (Figure 2.11 (c)), the function of the oxide is isolate the piezoresistors. The etching control is performed following the Arrhenius equation, which allows the calculation of the etching rate of Si as a function of the solution temperature [57].

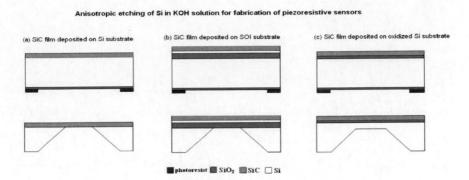

Figure 2.11. Illustration of the etching process of Si in KOH solution to form diaphragms: (a) Si substrate, (b) SOI substrate and (c) oxidized Si substrate.

Chapter 3

DESCRIPTION OF EXPERIMENTAL PROCEDURES: DEPOSITION, CHARACTERIZATION AND ETCHING

In this chapter are described the experimental procedures carried out to produce a-SiC thin films by PECVD and RF magnetron sputtering. It are also described the thermal annealing process and the etching conditions of the a-SiC films produced.

3.1. DEPOSITION OF SiC FILMS BY PECVD

3.1.1. Description of the PECVD Reactor Used

The PECVD system used consists of a cylindrical stainless steel deposition chamber with a total volume of 25 L. The vacuum system consists of two pumps, a diffuser and a mechanical, that are connected to the deposition chamber, which allows to obtain a background pressures of about 10^{-6} Torr. The gas flow is regulated by mass flow controllers calibrated for each type of gas used.

a

b

Figure 3.1. PECVD system: (a) general picture of the system and (b) inside the deposition chamber.

This PECVD system also possesses Pirani capacitance diaphragm gauges, RF generators and an RF amplifier. The substrates are placed on a stainless steel cathode (65 mm in diameter) cooled water that enables the deposition of SiC films practically at room temperature. The cathode is connected to a RF source (13.56 MHz) through an impedance matching. Figure 3.1 shows photographs of the PECVD system used.

3.1.2. PECVD Deposition Conditions

Amorphous silicon carbide thin films have been produced on p-type (100) silicon substrate using a PECVD system equipped with a radio frequency (RF) source operated at a frequency of 13.56MHz and power of 200W. The deposition conditions of five samples of SiC films, four un-doped and one nitrogen-doped, are showed in Table 3.1. All samples were produced under low SiH_4 flow ("silane starving plasma") [58]. In order to produce un-doped SiC films with different chemical composition, all the deposition parameters were kept constant and only the SiH_4 flow was varied. The doping of SiC film was performed by the introduction of N_2 gas during the deposition process.

Table 3.1. Deposition conditions of SiC films by PECVD

Sample	SiH_4 flow (sccm)	CH_4 flow (sccm)	Ar flow (sccm)	N_2 flow (sccm)	Pressure (Torr)	Deposition time (min)
P1	1.0	20	20	-	0.2	20
P2	2.0	20	20	-	0.2	20
P3	3.0	20	20	-	0.2	20
P4	4.0	20	20	-	0.2	20
P5	4.0	20	20	2.0	0.2	20

3.2. DEPOSITION OF SiC FILMS BY RF MAGNETRON SPUTTERING

3.2.1. Description of the Magnetron Sputtering Reactor Used

The RF magnetron sputtering reactor consists of a stainless steel chamber in cylindrical shape with 60 cm in height and 90 cm in diameter. The vacuum system consists of two pumps: one mechanical and the other is a diffuser. The electrical system is equipped with a radio frequency (RF) source operated at a frequency of 13.56 MHz. An impedance matching is used to ensure maximum efficiency of the RF source. The inside of the reactor consists of the magnetron cathode, the substrate holder, shutter and heating substrate system. The SiC (99.5% purity) target with 184 mm in diameter and 6mm thick is placed on the cathode. In Figure 4.2 are shown photographs of the RF magnetron sputtering system used. The Baratron pressure gauges, Ion gauge controller, gas flow and a liquid nitrogen trap also are part of the experimental apparatus. The liquid nitrogen trap has the function of preventing oil vapor molecules from migrating into the deposition chamber and affect the purity of the film.

a

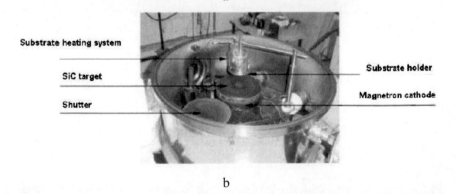

b

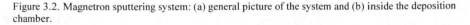

Figure 3.2. Magnetron sputtering system: (a) general picture of the system and (b) inside the deposition chamber.

3.2.2. Magnetron Sputtering Deposition Conditions

Table 3.2 shows the deposition conditions of six samples of SiC films by RF magnetron sputtering. In order to study the influence of the nitrogen concentration on film characteristics, all depositions parameters were kept constant and only the N_2 flow was varied.

Table 3.2. Deposition conditions of SiC films by RF magnetron sputtering

Sample	RF power (W)	Ar flow (sccm)	N_2 flow (sccm)	Pressure (10^{-3} Torr)	Deposition time (min)
M0	200	7.0	-	2.0	120
M1	200	7.0	0.7	2.0	120
M2	200	7.0	1.4	2.0	120
M3	200	7.0	2.1	2.0	120
M4	200	7.0	2.8	2.0	120
M5	200	7.0	3.5	2.0	120

3.3. POST-DEPOSITION THERMAL ANNEALING OF SiC FILMS

The techniques of RF magnetron sputtering and PECVD involve low temperatures, so it is expected that films deposited by these techniques are amorphous. In order to evaluate the effect of thermal annealing on the properties of these films, all samples produced in this study were subjected to an annealing process at 1000 °C in argon atmosphere for 1h.

3.4. REACTIVE ION ETCHING (RIE) OF SiC FILMS

Plasma etching studies were performed in a RIE parallel plate reactor using SF_6 as the reactive fluorinated gas and O_2 as an additive. The RIE system used consists of an aluminum cylindrical chamber with diameter of 230 mm and height of 130 mm, an RF source (13.56 MHz), a gas injection system and pressure control. A mass spectrometer with 1 atomic mass unit of resolution is connected to the RIE system. The plasma species are collected through a micro-orifice located at the entrance of the spectrometer. Figure 3.3 shows photographs of the RIE system.

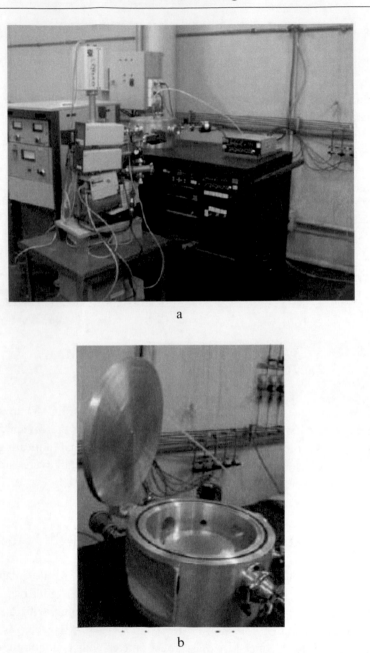

a

b

Figure 3.3. RIE system: (a) general picture of the system and (b) inside the etching chamber.

In Table 3.3 are showed the etching conditions of SiC films by RIE. The effect of O_2 concentration on the etching rate of the SiC films was studied for two concentrations, 20% and 80%, in SF_6 + O_2 gas mixture. To determine the etching rate was protected a small area of the sample. The etched thickness was measured by profilometry at three different points.

Table 3.3. Etching parameters of SiC films

Parameter	Value
Substrate temperature (°C)	20
Pressure (mTorr)	25
Power (W)	50
Time (min)	3
Gases total flow (sccm)	12

Chapter 4

ANALYSIS OF EXPERIMENTAL RESULTS

In this chapter are presented the chemical, structural, morphological, electrical, mechanical, and optical properties of the a-SiC films produced by PECVD and RF magnetron sputtering techniques. Etching studies of the a-SiC films are also presented.

4.1. PECVD SiC FILMS

4.1.1. Chemical Composition and Thickness

RBS measurements were performed to determine the chemical composition and thickness of PECVD SiC films. The concentration of species in the samples was calculated by means of computer simulations with the RUMP program. Table 4.1 summarizes RBS results and the RBS spectra are shown in Figure 4.1. It was observed that the films produced are non-stoichiometric silicon carbide (Si_xC_y) films with high carbon concentration and that the increase of SiH_4 flow in the deposition process promotes a decrease of the carbon content in the film. This reduction of carbon content is followed by the increase of the silicon concentration, promoting the formation of Si-C bonds. Also, it is possible to observe the presence of small amounts of oxygen for all analyzed films ($\sim$ 5%). The presence of this contaminant might have occurred either by the exposure of the samples to air or by oxygen contamination of the PECVD system. The sample P5 deposited with the addition of N_2 to SiH_4 / CH_4 / Ar gas mixture showed the following composition: 31% Si, 56% C, 7% N and 3% O.

Table 4.1. Composition, thickness and deposition rate of PECVD Si_xC_y films

Sample	SiH$_4$ flow (sccm)	Si (at.%)	C (at.%)	O (at.%)	Thickness (nm)
P1	1.0	9.0	82.0	5.0	500
P2	2.0	14.0	76.0	4.5	580
P3	3.0	18.0	73.0	5.0	640
P4	4.0	25.0	68.0	4.8	720

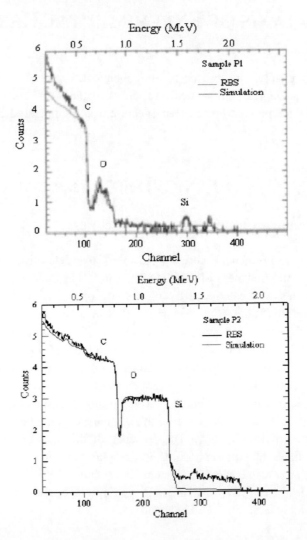

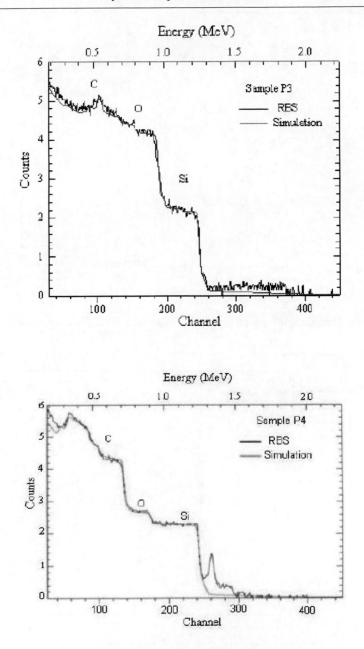

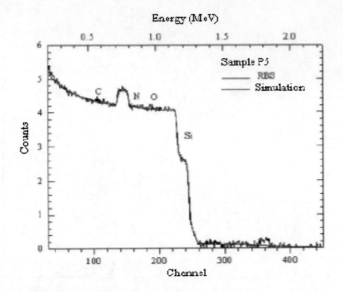

Figure 4.1. RBS spectra of the PECVD Si_xC_y films deposited under different SiH_4 flows.

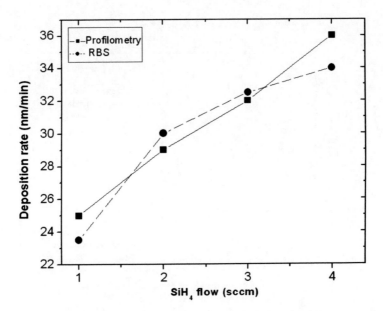

Figure 4.2. Deposition rate as a function of SiH_4 flow in PECVD deposition process.

The thicknesses of the films also were measured by profilometry. The deposition rate increases proportionally with the increase of SiH_4 flow (see Figure 4.2). This behavior is correlated with the increase of the silicon concentration in the gas feed which results in an increase of the number of atoms incorporated in the film [59].

4.1.2. Chemical Bonding and Structure of As-Deposited and Annealed SiC Films

Figure 4.3 shows (a) the infrared transmission spectra of PECVD Si_xC_y films deposited at different SiH_4/CH_4 flow ratios. All samples present stretching mode at ~610 cm^{-1}, ~820 cm^{-1} and ~1108 cm^{-1} that are corresponding to Si-H, Si-C and Si–O bonds respectively. The intensity of the Si-C stretching mode increases continuously with the increase of the SiH_4/CH_4 flow ratio. As can be observed in Figure 4.3 (b), the thermal annealing promotes structural modifications in the SiC films. The IR transmission bands observed are similar to the reported by other authors for silicon-carbon alloys [60,61].

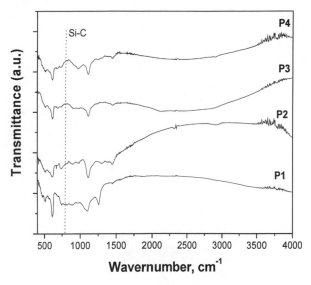

a

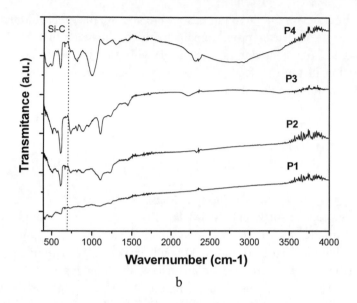

Figure 4.3. IR transmission spectra of PECVD Si_xC_y films: (a) as-deposited and (b) post-annealed.

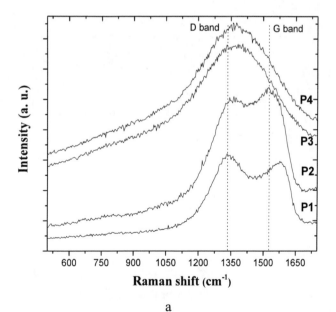

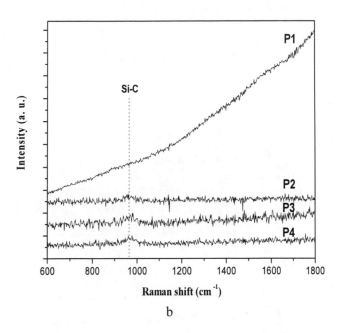

Figure 4.4. Raman spectra of PECVD Si_xC_y films: (a) as-deposited and (b) post-annealed.

Additional informations about chemical bonds were obtained from Raman spectra of as-deposited and annealed Si_xC_y films shown in Figure 4.4 (a) and (b) respectively. It has been observed that the as-deposited films do not exhibit peaks or bands corresponding to Si–C bonds. However, the a-Si_xC_y films deposited under SiH_4/CH_4 flow ratios of 0.05 and 0.1 evidence peaks associated to C–C bonds at ~1340 cm^{-1} (D band) and ~1585 cm^{-1} (G band). For SiH_4/CH_4 ratios most that 0.1, only the D band is observed. This occurs due to reduction of carbon concentration in the film.

After annealing, Raman spectra show that the films do not evidence D and G bands associated to C–C bonds. However, it was observed a weak peak corresponding to SiC phase. This indicates that the thermal annealing must be hindering the formation of C–C bonds and promoting the formation of the Si–C bonds.

The structural characterization was performed by XRD measurements. Figure 4.5 (a) shows the XRD spectra for the as-deposited films. As can be observed, the SiC films do not present diffraction peaks, which confirm their amorphous condition.

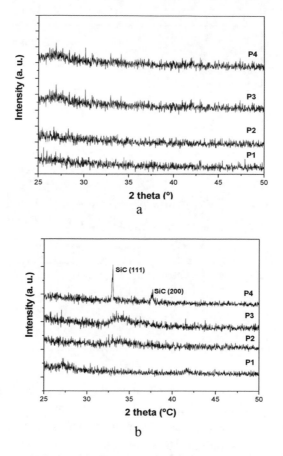

Figure 4.5. XRD spectra of PECVD Si_xC_y films: (a) as-deposited and (b) post-annealed.

It was also observed that after thermal annealing only the sample P4 with less carbon concentration showed visible peaks at about 33° and 37° which represent the indices of (1 1 1) and (2 0 0) β-SiC (see Figure 4.5 (b)). This result indicates that the SiC thin films with high carbon content exhibit a low crystallization degree.

4.1.3. Morphology and Roughness

The morphology of the films was investigated by AFM, and the parameter analyzed was the RMS roughness obtained from areas of 1.0μm x 1.0μm. Figure 4.6 shows the AFM image of the PECVD Si_xC_y film (sample P4).

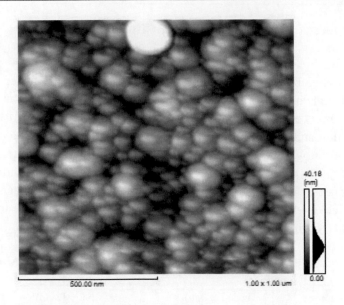

Figure 4.6. AFM image of the PECVD Si_xC_y film (sample P4).

As can be observed in Figure 4.7, the RMS roughness of the film depends on the concentration of SiH_4 used in the deposition process. It is observed that the roughness of the PECVD Si_xC_y film tends to increase with Si incorporation in the film increases.

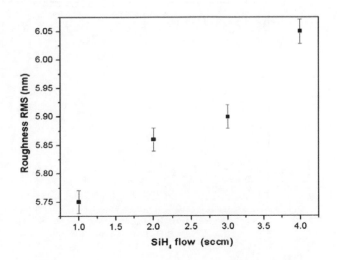

Figure 4.7. Roughness of the PECVD Si_xC_y films deposited under different SiH_4 flow.

4.1.4. Electrical Resistivity and Elastic Modulus

The resistivity of PECVD Si_xC_y films was determined by four-point technique. Figure 4.8 shows the sheet resistance of the films, as-deposited and after thermal annealing process, as a function of SiH_4 flows used in the deposition process. It is observed that all samples, independent of its chemical composition and heat treatment, showed a sheet resistance of about 10^5 $\Omega/\square$. However, nitrogen-doped Si_xC_y film (sample P5) has a sheet resistance of 2.7 x 10^2 $\Omega/\square$. This shows that the in situ doping by the addition of nitrogen was efficient because reduced the sheet resistance of three orders of magnitude.

In Table 4.2 are shown the resistivity of the films and the thicknesses measured by profilometry.

Table 4.2. Electrical resistivity and thickness of the PECVD Si_xC_y films

Sample	Thickness (nm)	Resistivity (Ω.cm)
P1	500	12.5
P2	580	10.4
P3	640	12.8
P4	720	12.3
P5	480	1.3×10^{-2}

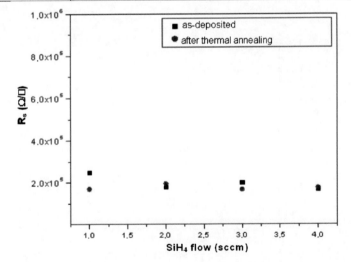

Figure 4.8. Sheet resistance of PECVD Si_xC_y films deposited on thermally oxidized Si substrates at different SiH_4 flow rates.

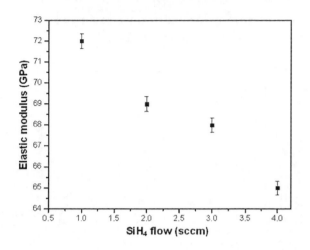

Figure 4.9. Elastic modulus of PECVD Si$_x$C$_y$ films deposited on thermally oxidized Si substrates at different SiH$_4$ flow rates.

The elastic modulus of the PECVD Si$_x$C$_y$ films was determined by nanoindentation. The samples were measured at an indentation depth equal to 10% of film thickness. The results analyzed by the method of Oliver and Pharr are shown in Figure 4.9. It was observed that the elastic modulus of the films decreased from 72 GPa to 65 GPa with the increase of SiH$_4$ flow in deposition process. This occurs because the mechanical properties of carbon-based films depend mainly on the strength of C-C bonds.

4.1.5. Etching Studies

PECVD Si$_x$C$_y$ films were etched in a RIE system at different O$_2$ concentrations in the SF$_6$/O$_2$ gas mixture: 20% and 80%. The variation of O$_2$ concentration allowed evaluating the role of the carbon content in film and the influence of the oxygen addition gas on the etch rate and morphology of the etched material. The etching rate of the SiC films as a function of carbon content for different O$_2$ concentrations in the SF$_6$/O$_2$ gas mixture is shown in Figure 4.10. For both oxygen concentrations, the etch rate decreases with the increases of carbon content in SiC film. Figure 4.10 also shows the etch rate increases with the increases of the O$_2$ concentration from 20 to 80%. This occurs because the addition of oxygen in fluorinated plasmas enables increase of the fluorite ion generation enhancing the etching of the silicon by the formation of SiF_4 volatile specie and also the etching of rich carbon layers by the formation of CO and CO_2 volatile species [62].

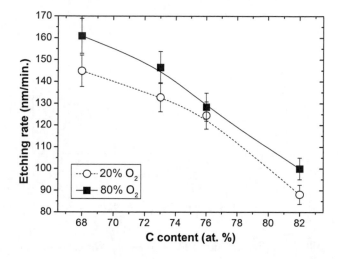

Figure 4.10. Etching rate of the PECVD Si_xC_y films as a function of carbon content for different O_2 concentrations in SF_6/O_2 gas mixture.

The roughness and morphology of the PECVD Si_xC_y films after reactive ion etching in SF_6/O_2 plasmas were investigated by AFM. The RMS roughness of the as-deposited films varied between 5.75 and 6.0 nm. When the film is submitted to the etching process, it was observed a decrease in RMS roughness. This reduction is dependent of the O_2 concentration, for 20% O_2 was observed the lesser RMS roughness values, of the order of 0.75 nm, while for 80% O_2 had been measured values of up to 1.8 nm. In order to illustrate this effect, Figure 4.11 shows the surface morphology of the SiC film obtained at SiH_4 = 4.0 sccm for conditions: (a) after etching (80% O_2) and (b) after etching (20% O_2). A surface smoother and free of defects is showed for the case of 20% O_2.

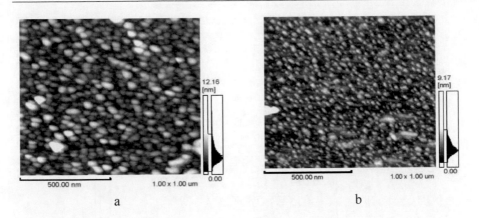

Figure 4.11. AFM dynamic mode image of the surface of PECVD Si_xC_y film for carbon concentration = 68%: (a) after etching (80% O_2) and (b) after etching (20% O_2). The transverse dimension of the scanned area is 1.0 ×1.0 μm.

The mass spectra obtained during the etching of Si_xC_y film in an RIE system is showed in Figure 4.12 and illustrate three different processes: the effluent of pure SF_6 gas with discharge (Figure 4.12 (a)) and without discharge (Figure 4.12 (b)), and the effluent generated in a mixture of SF_6 (50%)+O_2(50%) gas discharge (Figure 4.12 (c)). In the absence of discharge, it was observed that the most peaks resultant is fragmentation of SF_6 gas. Other peaks of lesser intensity as H_2^+ (mass 2), H_2O^+ (mass 18) and CO^+ (mass 28), can also be observed. When the discharge is working, the RIE process promotes the increase of fluorine atoms represented in mass spectra by F^+ peak (mass 19), and others fluorine-contain peaks as HF^+ (mass 20) and SiF_3^+ (mass 85), where the latter represents the Si etching volatile product, SiF_4. As it is known, during a pure SF_6 etching the fluorine atoms react preferentially with Si atoms in the first layers of the SiC crystal, and the ion bombardment partially breaks Si–C bonds liberating the volatile product SiF_4. However, this preferential Si etch results in the formation of a carbon-rich layer that is a rate-limiting step for SiC etching. Moreover, the break of Si–C bonds facilitates the recombination of carbon atoms that, as observed in the Figure 4.12 (b), reacts with fluorine atoms forming the ultimate-carbon containing etch products CF^+ (mass 31), CF_2^+ (mass 50) and CF_3^+ (mass 69) in a pure SF_6 discharge [63].

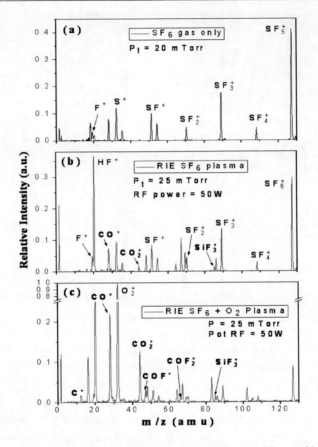

Figure 4.12. Mass spectra of the effluent from reactor in the conditions: (a) SF_6 gas without discharge, (b) SF_6 gas with discharge, and (c) SF_6 (50%) + O_2 (50%) gas mixture with discharge. Fragmentation pattern for SF_6 gas: $SF_5^+ = 100$, $SF_4^+ = 10$, $SF_3^+ = 35$, $SF_2^+ = 10$, $SF^+ = 22$, $F^+ = 1$. The intensity was normalized to 1 for maximum value [63].

As shown in the Figure 4.12(c), with the insertion of O_2 in the SF_6 discharge, a significant increase in primary carbon etching products CO^+ (mass 28) and CO_2^+ (mass 44) can be observed. Additionally, it is possible to note the appearance of secondary products; namely, COF^+ (mass 47) and COF_2^+ (mass 66). Thus, the removal of carbon atoms in SiC film is enhanced and, as a result, the etching rate increases.

As observed from literature [64], in the case of Si etching, the main consequence of O_2 addition is the increase of the F atomic concentration in the etching gas mixture by the inhibition of the recombination reaction of the SF_6 that is described by the equation:

$$SF_{x-1} + \left\{ \begin{array}{c} F \\ F_2 \end{array} \right\} \rightarrow SF_x (+ F) \; (x = 4\text{-}6) \tag{4.1}$$

In which, the following reactions given for the equations (4.2) and (4.3) are derived:

$$O + SOF_3 \rightarrow SO_2F_2 + F \tag{4.2}$$

$$e + SOF_4 \rightarrow SOF_3 + F^- \tag{4.3}$$

Consequently, increasing the Si etch until an optimal value of O_2 concentration.

However, after this optimal value the increase of O_2 concentration in the mixture, promotes the oxidation of the SOF_2 molecule, occurring the reaction:

$$O + SOF_2 \rightarrow SO_2F_2 \tag{4.4}$$

This reaction is favored in relation to the equations (4.2) and (4.3), reducing the density of F radicals and consequently the Si etching. For SiC etching, this phenomenon also occurs and could be verified in our experiments through the monitoring of the species SiF_3^+ and F^+ as a function of $O_2\%$ (see Figure 4.13 (a)). When O_2 ratio exceeds approximately 20% the peaks of SiF_3^+ and F^+ diminishes, following, a similar behavior of the etch rate illustrated in Figure 4.10. As can be observed from Figure 4.13(b) a high increase in the concentration of primary volatile etch products of the carbon, caused by the increase of the dissociation of the species with C-F bonds. An example of this behavior is the reduction of the CF_x^+ peaks, as presented in the spectrum of the Fig. 4.12(c).

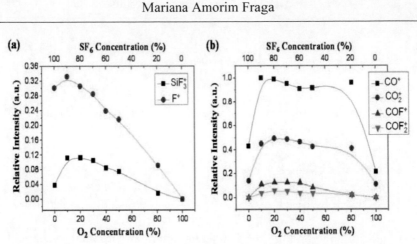

Figure 4.13. The normalized intensity of the SiF_3^+, F^+ (a) and CO^+, CO_2^+, COF^+, COF_2^+ (b) as a function of O_2 concentration in $SF_6 + O_2$ gas mixture. The intensity was normalized to 1 for maximum value.

Table 4.3 compares the etching rates values obtained for different O_2 concentrations. It is observed that after annealing the etching rate of the film suffered a significant reduction caused by the structural order of the Si-C bonds that makes the film more resistant to the etching process.

Table 4.3. Comparison between the etching rates of PECVD Si_xC_y film (with 68% carbon) as-deposited and after thermal annealing

	Etching rate (nm/min)	
O_2 concentration in SF_6/O_2 gas mixture (%)	as-deposited	After thermal annealing
20	145	30
80	160	12,5

4.1.6. Optical Characteristics

The experimental transmission and reflection spectroscopy analysis of the films was performed using a halogen lamp as the light source. The light beam was diffracted by a plane diffraction gratings attached to a step motor. Both, reflection and transmission signals are acquired simultaneously. The spectroscopic characterization of PECVD Si_xC_y thin film (sample P4) is showed in Figure 4.14. It was observed that Si_xC_y film presents good transmittance.

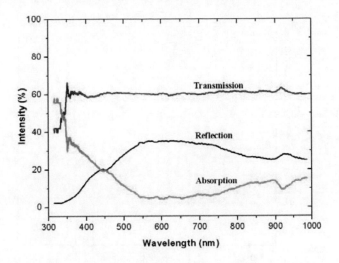

Figure 4.14. The spectroscopic characterization of PECVD Si_xC_y thin film (sample P4).

In Figure 4.15 is shown the absorption spectrum of the sample P4 and Figure 4.16 shows the optical gap energy as a function of SiH_4 flow during deposition process of the PECVD Si_xC_y films. As the electrical resistivity, the optical gap energy did not varied significantly in function of the reduction of carbon content in the film.

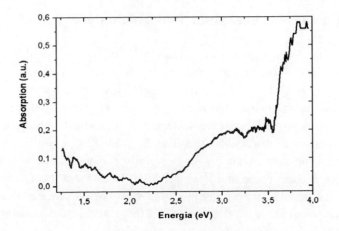

Figure 4.15. Absorption spectrum of PECVD Si_xC_y thin film (sample P4).

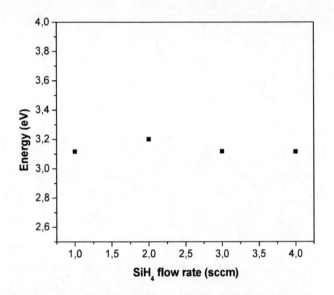

Figure 4.16. Optical gap energy of the PECVD Si_xC_y thin films as a function of SiH_4 flow during deposition process.

4.2. SPUTTERED SiC FILMS

4.2.1. Chemical Composition and Thickness

In Table 4.4 are shown the RBS results of the sputtered SiC films and Figure 4.17 shows the spectra obtained. It was observed that the film deposited without introduction of nitrogen (sample M0) is a stoichiometric compound of silicon and carbon. It was observed that the introduction of N_2 reactive gas in the deposition process promotes a decrease in Si and C atom concentrations due to insertion of nitrogen in the film. Sequentially, it can be observed that the gradual increase of the N_2/Ar flow ratio promotes a reduction in the Si content to a fixed value and a progressive decrease in the C content of the films. In all samples, it was possible to identify the presence of small amounts of oxygen for all analyzed films ($\sim 2\%$). The presence of this contaminant may be caused by the surface oxidization of the films.

Table 4.4. Film composition measured by RBS

N₂/Ar flow ratios	Sample	Si (at. %)	C (at. %)	N (at. %)	O (at. %)
0	M0	48	48	-	2.0
0.1	M1	28	24	46	1.5
0.2	M2	28	22	48	1.5
0.3	M3	28	21	49	2.0
0.4	M4	28	16	53	2.0
0.5	M5	25	17	56	1.5

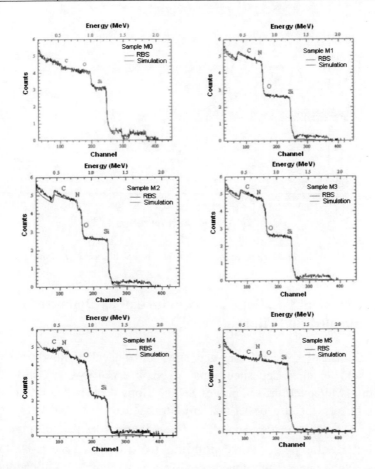

Figure 4.17. RBS spectra of the sputtered sputtered SiC and SiC$_x$N$_y$ films deposited under different N₂/Ar flow ratios.

The growth rate of the sputtered films was determined as a function of N_2/Ar flow ratio by RBS and profilometry. The increase in N_2 flow rate from 0 to 60 sccm affects the growth rate as shows Figure 4.18. It is observed that when the concentration of N_2 gas increases the film thickness tends to decrease. This result shows that the sputtering yield of SiC target decreases when nitrogen is introduced to deposition chamber [65].

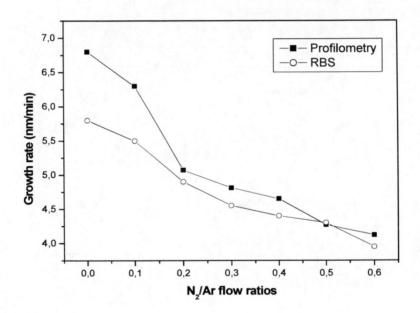

Figure 4.18. Growth rate of the sputtered SiC and SiC_xN_y films at different N_2/Ar flow ratios.

4.2.2. Chemical Bonding and Structure of as-Deposited and Annealed SiC Films

The infrared spectra for as-deposited and annealed films are shown in Figure 4.19. In all, as-deposited films can be observed Si–C stretching mode at 841 cm^{-1}. Moreover, the samples of SiC_xN_y films deposited at different N_2/Ar flow ratio present also peaks that are attributed to Si-N and C-N stretching modes. The spectra after annealing (see Figure 4.19 (b)) change in the position of the peaks corresponding to Si-C, Si-N and C-N stretching modes, which indicate that this process caused structural modifications in the films.

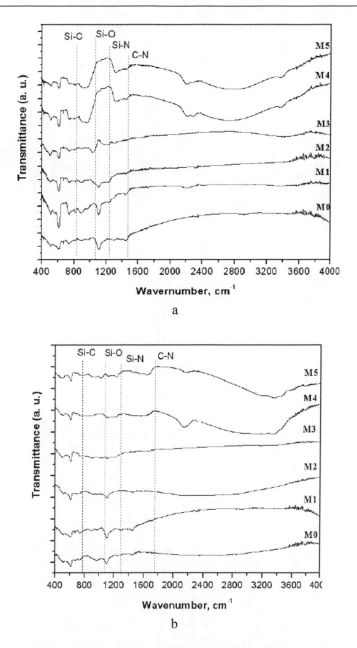

Figure 4.19. IR transmission spectra of sputtered sputtered SiC and SiC_xN_y films: (a) as-deposited and (b) post-annealed.

The Raman spectra of as-deposited SiC_xN_y films (see Figure 4.20 (a)) consist of two broad bands centered around at 1350 and 1580 cm-1 characteristic of the presence of amorphous carbon. Corresponding to sp^2 (graphite) and sp^3 (diamond) respectively. However, the Raman spectrum of the SiC film (sample M0) does not present evidences of any peak corresponding to C-C bonds.

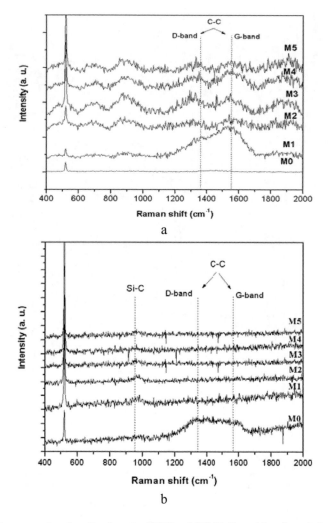

Figure 4.20. Raman spectra of sputtered sputtered SiC and SiC$_x$N$_y$ films: (a) as-deposited and (b) post-annealed.

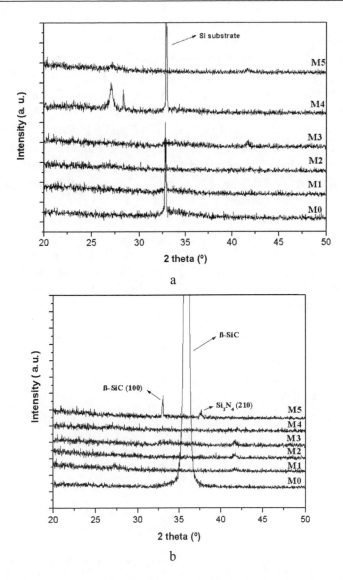

Figure 4.21. XRD spectra of sputtered sputtered SiC and SiC$_x$N$_y$ films: (a) as-deposited and (b) post-annealed.

It is also observed that all the films deposited (with or without nitrogen in the gas mixtures) present a relatively weaker band around 830 cm^{-1}, which is typical of amorphous SiC structure and a band around 500 cm^{-1} which is due to amorphous silicon. After annealing (Figure 4.20 (b)), the spectrum of the

SiC film (sample M0) shows the D and G bands associated with the C-C bonds and the SiC_xN_y films showed a weak signal centered at 930 cm^{-1} related Si-C.

The amorphous nature of the as-deposited films was confirmed by XRD spectra. As can be observed in Figure 4.21 (a), the as-deposited films do not present any diffraction peaks. These results clearly indicate that the low temperature deposition process does not promote the formation of crystalline phases. However, the samples M0 (SiC film) and M5 (SiC_xN_y film) present peaks associated to Si substrate. Moreover, SiC_xN_y film (samples M3 and M4) exhibit one weak broad peak around 42° which characteristic of the SiO_2. This fact is confirmed by RBS and FTIR analyses that identified the presence of oxygen in the films deposited.

Figure 4.21(b) shows the XRD spectrum for annealed films. Indeed, crystallization occurred and, as can be observed, the sputtered SiC film presents a β-SiC (100) phase and, the SiC_xN_y film deposited at higher N_2/Ar flow ratio (sample M5), a β-SiC (111) indicating that the β-SiC phase orientation was affected by the nitrogen addition. Moreover, the sample M5 also presented the Si_3N_4 phase.

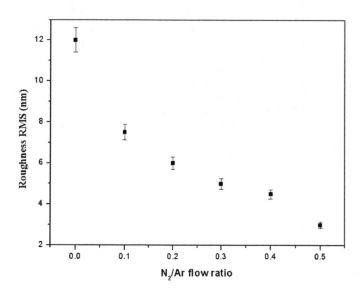

Figure 4.22. Roughness of the sputtered sputtered SiC and SiC_xN_y films deposited under different N_2/Ar flow ratios.

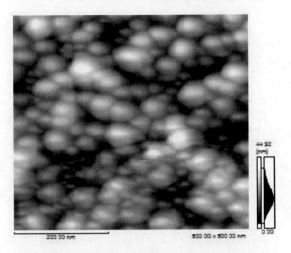

a

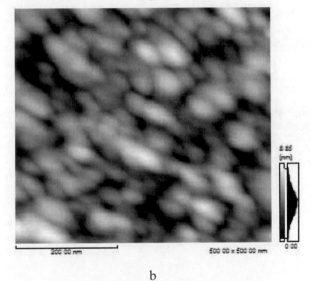

b

Figure 4.23. AFM images of the sputtered films: un-doped SiC film (sample M0) and (b) SiC_xN_y film (sample M3).

4.2.3. Morphology and Roughness

Figure 4.22 shows that the addition of nitrogen in the sputtering process promotes a reduction in the RMS roughness. AFM images of the films (see

Figure 4.23) indicate that the SiC_xN_y films are much smoother than SiC film. This can be understood by the fact that a reactive process forms the films. Therefore, higher nitrogen flow rate increases the reactive process allowing the formation of denser films with fewer voids.

4.2.4. Electrical Resistivity and Elastic Modulus

It was not possible to determine the resistivity of the as-deposited SiC_xN_y films with four-points probe system used. After the thermal annealing process, the sheet resistances of the films are shown in Figure 4.24 and the resistivities are shown in Table 4.5. Note that the introduction of nitrogen to the sputtering process caused an increase in resistivity up to an order of magnitude. This result is similar to those reported by other authors [66,67] and indicates that in these deposition conditions the nitrogen does not act as dopant and promotes the formation of SiC_xN_y ternary compound.

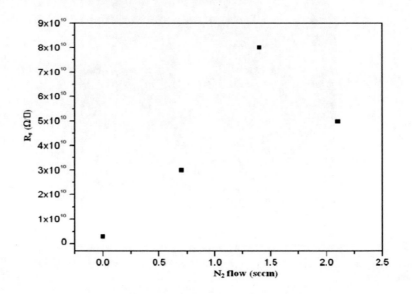

Figure 4.24. Sheet resistance of sputtered sputtered SiC and SiC_xN_y films deposited deposited on thermally oxidized Si substrates at different N_2 flow rates.

**Table 4.5. Electrical resistivity and
thickness of the sputtered SiC films**

Sample	Thickness (nm)	Resistivity (MΩ.cm)
M0	816	0.25
M1	756	2.27
M2	608	4.87
M3	577	2.9

Figure 4.25 shows the elastic modulus and hardness values of the a-SiC and a-SiC$_x$N$_y$ films as a function of indentation depth. The values obtained indicate that the hardness and reduced elastic modulus of the films increase with nitrogen flow rate increases. This increase is due to formation of C-N and Si-N phases in the films [68].

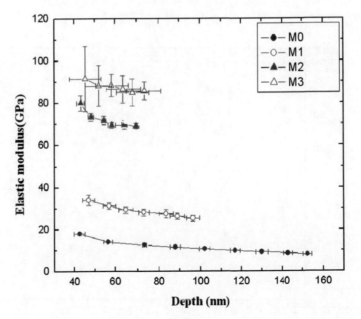

a

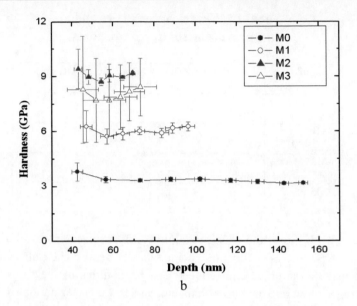

b

Figure 4.25. Elastic modulus (a) and hardness (b) of the sputtered SiC and SiC_xN_y films deposited on thermally oxidized Si substrates measured by nanoindentation.

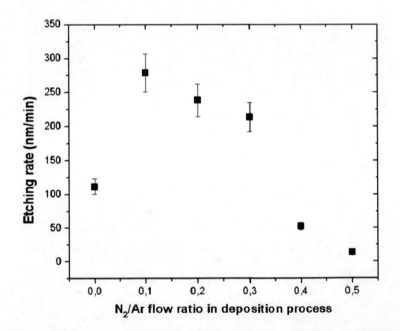

Figure 4.26. Etch rates of a-SiC and a-SiC_xN_y films as a function of N_2/Ar flow ratio in deposition process.

4.2.5. Etching Studies

The sputtered a-SiC and a-SiC$_x$N$_y$ films have been etched in a SF$_6$ (80%) + O$_2$ (20%) plasma environment using RIE technique. In Figure 4.26, the dependence of the film etch rates on nitrogen concentration, used in deposition process are plotted. It can be observed that the a-SiC$_x$N$_y$ etch rates tend to be higher as compared with a-SiC etch rate. This is not true when the nitrogen content in film is greater than 50 at.%. This occurs because the low concentration of C atoms in the film providing links between Si and N probably forming the Si$_3$N$_4$ compound, which is more resistant to the etching process.

To better understand the etching results, a quadruple mass spectrometry analysis of effluents generated during etching process was performed. Initially, a temporal monitoring of the main species formed during the etching process was done. When the discharge is on, the RIE process promotes the increase of fluorine atoms represented in mass spectra by F$^+$ peak (mass 19), and others fluorine-contain peaks as HF$^+$ (mass 20) and SiF$_3$$^+$ (mass 85), where the latter represents the Si etching volatile product, SiF$_4$. As it is known, during a pure SF$_6$ etching the generated fluorine atoms react preferentially with Si atoms in the first layers of the SiC crystal, and the ion bombardment partially breaks Si–C bonds liberating the volatile product SiF$_4$. However, this preferential Si etching results in the formation of a carbon-rich layer that is a rate-limiting step for SiC [69]. Moreover, the break of Si–C bonds facilitates the recombination of carbon atoms that reacts with fluorine atoms forming the ultimate-carbon containing etch products CF$^+$ (mass 31), CF$_2$$^+$ (mass 50) and CF$_3$$^+$ (mass 69) in a pure SF$_6$ discharge, have low contribution in etch process. Adding O$_2$ in the SF$_6$ discharges a significant increase in primary carbon etching products - CO$^+$ (mass 28) and CO$_2$$^+$ (mass 44) - is observed in the mass spectrum. Additionally, it is possible to note the appearance of secondary products; namely, COF$^+$ (mass 47) and COF$_2$$^+$ (mass 66). Thus, the removal of carbon atoms in SiC film is enhanced, and, as a result, the etching rate increases.

In Figure 4.27 is illustrated the variation of the main products of etching (SiF$_3$$^+$, CO$^+$ and CO$_2$$^+$) and the main species of etchant gases (O$_2$$^+$ and SF$_5$$^+$) during the etching of the sample M1. It can be observed that when the discharge is on, a significant increase in the partial pressure of CO$^+$ and CO$_2$$^+$ species can be observed, as well as SiF$_3$$^+$. However, there is a decline of the O$_2$$^+$ and SF$_5$$^+$ species indicating their fragmentation or recombination with the substrate. Still, from this figure is possible to verify that the CO$^+$ specie presents a higher variation in partial pressure compared with the other species examined. Relating to etching process of silicon atom, the species with higher signal is the SiF_3^+. Based on these results, a comparison of the temporal variation of species CO^+ and SiF_3^+ was done for a-SiC_xN_y samples. Figure 4.28 shows that the signal of CO^+ decreases with N_2/Ar flow ratio. In particular, for the M4 and M5 samples the CO^+ partial pressures are much lower than for the other samples, indicating a low etching of the substrate material. Although not illustrated here, it was observed the appearance of the SiF_3^+ specie during all etching processes, however, the partial pressure throughout the process was equal for all samples during the etching period (around 9.5×10^{-8} Torr). This fact shows that the reaction between C and O atoms is prevailing on the etching process of SiC_xN_y film.

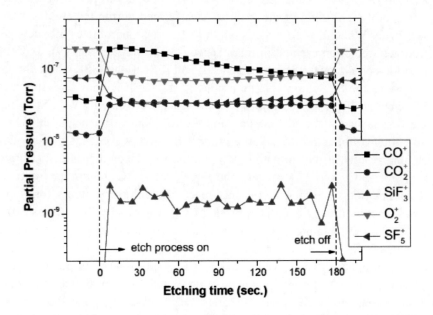

Figure 4.27. Time evolution of monitored species CO^+, CO_2^+, SiF_3^+, O_2^+ and SF_5^+ during the etching process of a-SiC_xN_y film (sample M1). The partial pressure is relative the total pressure ($\sim 2 \times 10^{-5}$ Torr) in the vacuum system of the mass spectrometer [70].

Therefore, from the results presented some aspects about the process may be proposed. When there is nitrogen content < 50at.% the SiC_xN_y film becomes more susceptible to the etching process, perhaps because of the decline of Si–C bonds and increase of C–N bonds. With the reduction of the carbon concentration in the film, the nitrogen atoms tend to form more links with the silicon atoms, providing a formation of a SiC_xN_y film with large content of silicon nitride.

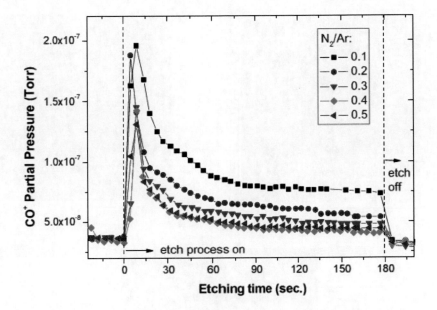

Figure 4.28. Time evolution of CO^+ specie during the etching process of a-SiC_xN_y films [70].

Using an AFM, the morphology and roughness of the reactive-ion etched samples were investigated. AFM images are shown in Figure 4.29. The etched sample presented RMS roughness between 3.5 and 5.5 nm.

4.2.6. Optical Characteristics

The optical band-gap of the SiC and SiC_xN_y thin films deposited by RF magnetron sputtering were determined from the absorption spectra using a similar procedure to that described for the PECVD Si_xC_y films (section 4.1.6). Four samples were analyzed: M0 (stoichiometric SiC film) and three samples of SiC_xN_y films with different nitrogen content (M1, M3 and M5). In Figure 4.29 are shown the optical band-gap as a function of N_2/Ar flow ratio. As can be observed, the introduction of nitrogen in the deposition process did not change significantly the optical band-gap of the films. However, the literature shows that the nitrogen doping of SiC films causes a reduction in optical band-gap and electrical resistivity of this material. This result also indicates that in the sputtering deposition conditions used the nitrogen did not act as dopant and promoted the formation of the silicon carbonitride.

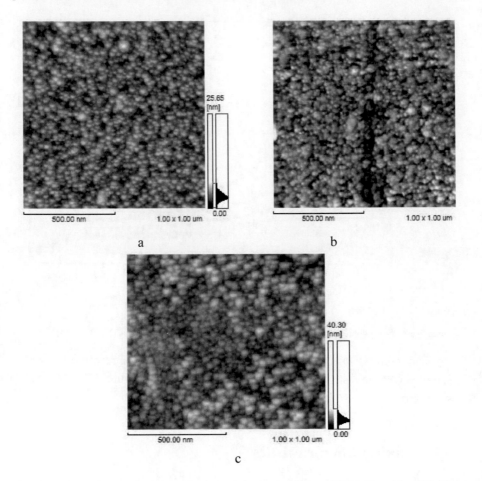

Figure 4.29. AFM dynamic mode image of the surface of etched SiC and SiC$_x$N$_y$ films: (a) sample M0, (b) sample M1 and (c) sample M2.

Figure 4.30 shows reflection spectra of the sputtered films. It was observed that the SiC film (sample M0) presents a behavior different of the SiC$_x$N$_y$ films. Table 4.6 compares the thickness of the films obtained by different three techniques: RBS, profilometry and multiple reflections. As can be seen the values are close indicating that any of these could be used to measure the thickness of films accurately.

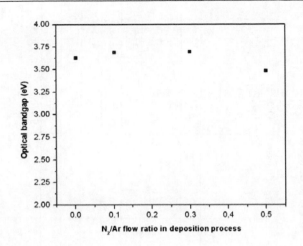

Figure 4.29. Optical bandgap of the sputtered SiC and SiC_xN_y films.

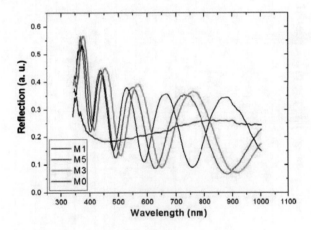

Figure 4.30. Reflection spectra of the sputtered SiC and SiC_xN_y films.

Table 4.6. Comparison between the thicknesses of the SiC and SiC_xN_y films measured by different techniques: multiple reflections, RBS and profilometry

	Thickness (nm)		
Sample	Reflection	Profilometry	RBS
M0	753	816	696
M1	606	608	660
M3	549	577	546
M5	441	494	475

4.3. COMPARISON BETWEEN THE PROPERTIES OF PECVD AND SPUTTERED SiC FILMS

Table 4.7. Comparison among the properties of the SiC films produced

Sample	Deposition process	N_2 content in deposition process (%)	Sheet resistance ($\Omega/\square$)	Sheet resistance after annealing ($\Omega/\square$)	Elastic modulus (GPa)	Roughness (nm)	Optical bandgap (eV)	Etching rate (nm/min)
P1	PECVD	-	2.5×10^5	1.7×10^5	72	5.75	3.117	85
P2	PECVD	-	1.75×10^5	1.9×10^5	69	5.86	3.2	130
P3	PECVD	-	2.0×10^5	1.65×10^5	68	5.9	3.118	135
P4	PECVD	-	1.7×10^5	1.7×10^5	65	6.5	3.116	145
P5	PECVD	5.0	2.7×10^2	1.8×10^2	57	6.0	2.9	138.7
M0	Sputtering	-	-	3.2×10^9	17	12	3.627	111.67
M1	Sputtering	10	-	4.8×10^{10}	28	7.5	3.688	278.66
M2	Sputtering	20	-	7.2×10^{10}	72	6.0	3.62	238.66
M3	Sputtering	30	-	6.7×10^{10}	88	5.0	3.686	53.33
M4	Sputtering	40	-	-	92	4.5	3.65	213.66
M5	Sputtering	50	-	-	90	3.0	3.48	17

Table 4.8. Comparison among the structural properties of the SiC films produced

Sample	Transmission FTIR spectra Bands observed		Raman spectra Bands observed		XRD spectra Peaks observed	
	as-deposited	annealed	as-deposited	annealed	as-deposited	annealed
P1	Si-H Si-C Si-O	Si-O	C-C (bands D and G)	-	-	-
P2	Si-H Si-C Si-O	Si-H$_2$ Si-C	C-C (bands D and G)	Si-C	-	-
P3	Si-H Si-C Si-O SiCH$_n$	Si-H$_2$ Si-C	C-C (band D)	Si-C	-	-
P4	Si-H Si-C Si-O SiCH$_n$	Si-H$_2$ Si-C	C-C (band D)	Si-C	-	SiC (111) SiC (200)
P5	Si-H Si-C Si-O Si-N	Si-C Si-O Si-N	-	Si-C	-	-
M0	Si-C Si-O	Si-C	-	C-C (bands D and G)	-	β-SiC (111)
M1	Si-C Si-O	Si-C	C-C (bands D and G)	Si-C	-	-

Sample	Transmission FTIR spectra		Raman spectra		XRD spectra	
	Bands observed		Bands observed		Peaks observed	
	as-deposited	annealed	as-deposited	annealed	as-deposited	annealed
M2	Si-C Si-O	Si-C Si-O	C-C (bands D and G)	Si-C	-	-
M3	Si-C Si-O Si-N C-N	Si-C Si-O Si-N C-N	C-C (bands D and G)	Si-C	-	-
M4	Si-C Si-O Si-N C-N	Si-C Si-O Si-N C-N	C-C (bands D and G)	Si-C	-	-
M5	Si-C Si-O Si-N C-N	Si-C Si-O Si-N C-N	C-C (bands D and G)	Si-C	-	β-SiC (100) Si_3N_4 (210)

Chapter 5

AN OVERVIEW ON SiC PIEZORESISTIVE SENSORS

In this chapter are presented a brief description of the piezoresistive effect and the evolution of the piezoresistive sensors based on SiC. Moreover, the experimental procedures to characterize the piezoresistive properties of SiC films are presented.

The last section of this chapter shows a methodology for the design, fabrication, packaging, and test of a prototype of piezoresistive pressure sensor based on a-SiC film.

5.1. THEORETICAL BACKGROUND

The piezoresistive effect provides an easy and direct energy/signal transduction mechanism between the mechanical and the electrical domains due to this is the sensing principle most common in micromachined sensors. Nowadays, piezoresistive sensors are used for a wide variety of applications including strain gauges, pressure sensors, accelerometers, and tactile sensors among others [71,72].

Lord Kelvin first reported the piezoresistive effect in 1856 [73]. This effect is characterized by changing resistivity of a material due to applied mechanical. The change in resistance is linearly related to the applied strain according to the equation below:

$$\frac{\Delta R}{R} = GF.\varepsilon \qquad\qquad (5.1)$$

Where GF is a proportional constant called gauge factor and ε is the strain.

There are two main types of piezoresistive materials: the metals and the semiconductors. Piezoresistive sensors based on thin-film metals do not compare favorably with semiconductors in terms of gauge factors, whereas the GF of silicon is more than 100 the of metals is less than 2.0.

Many of the commercial pressure, force and acceleration sensors are based on piezoresistive effect in silicon. The main drawback of these sensors is the low performance at high temperature [74]. This has motivated several researches on semiconductor materials, with chemical and mechanical stability, to substitute the silicon in the development of piezoresistive sensors for high temperature applications.

Among these materials, silicon carbide (SiC) is one of the most promising due to its excellent physical properties and compatibility with the microfabrication techniques based on silicon technology. In recent years, some works on the fabrication and characterization of piezoresistive sensors based on SiC have been reported.

In 1997, Okojie *et al* investigated the performance of a (6H)-SiC piezoresistive pressure sensor at temperature up to 500°C. The sensor developed consists of n-type 6H-SiC piezoresistors on a p-type 6H-SiC circular diaphragm. The tests performed showed that at room temperature the sensor has a full-scale output of 40.66 mV at 1000 psi decreasing to 20.3 mV at 500°C [75].

Also in 1997, Ziermann *et al* developed pressure sensors based on 3C-SiC films deposited on SOI substrates. This sensor consists four 3C-SiC piezoresistors on a SOI circular diaphragm. It was tested from room temperature up to 400°C and pressures up to 500 kPa [76].

In 2001, Wu *et al* reported a single crystalline 3C-SiC piezoresistive pressure sensor that exhibited a sensitivity of 0.1015 mV/V.psi at room temperature and 0.0534 mV/V.psi at 385°C [77].

In recent work [78], a piezoresistive pressure sensor based on PECVD a-SiC films was reported. The structure of the sensor consists of six a-SiC piezoresistors, configured in Wheatstone bridge, on a SiO_2/Si square diaphragm. The sensor exhibited a sensitivity of 0.33 mV/V.psi.

5.2. EXPERIMENTAL

5.2.1. Fabrication of Test Structures for Piezoresistive Measurements of SiC Films

In order to characterize the piezoresistive properties of a-SiC films produced by PECVD and sputtering, it were fabricated test structures that consist of one thin-film resistor with one Ti/Au electrical contact at each extreme of the resistor.

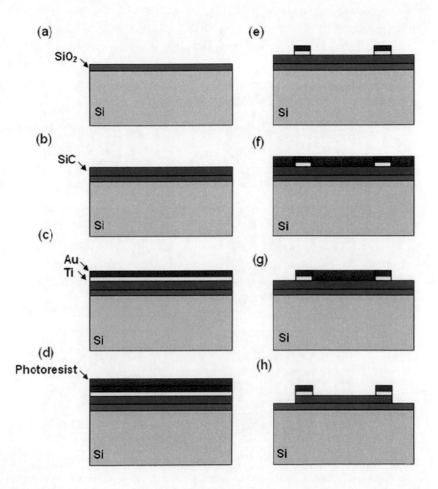

Figure 5.1. Schematic representation of fabrication of the test structures used to characterize the piezoresistive properties of the a-SiC films.

The fabrication process of the thin-film resistors is shown schematically in Figure 5.1 and can be described by the following steps:

a) Thermal oxidation of Si wafer;
b) Deposition of the SiC film;
c) Deposition of Ti and Au on the SiC film;
d) Photolithography to pattern the Ti/Au electrical contacts;
e) Unwanted Ti and Au were etched using etchant solution;
f) Photolithography to pattern the SiC thin-film resistors;
g) Reactive ion etching (RIE) of the regions of the SiC film do not protected by photoresist using SF_6/O_2 gas mixture;
h) Final structure of the SiC thin-film resistor.

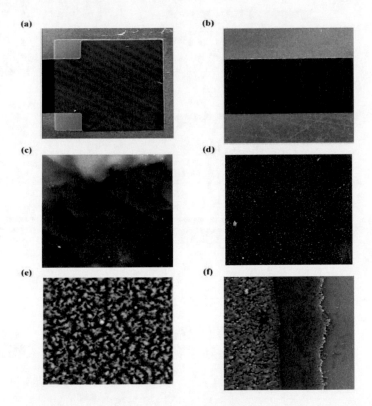

Figure 5.2. SEM images of the test structure fabricated: (a) Ti/Au electrical contact (100x), (b) PECVD SiC thin-film resistor (100x), (c) surface of the electrical contact (1000x), (d) surface of the electrical contact (5000x) and (e) PECVD SiC thin film (5000x) and (f) SiC / SiO_2/ Si interface (5000x).

5.2.2. Experimental Setup Used for Electrical and Piezoresistive Characterization of Test Structures

The I-V characteristics of the thin-films resistors produced were investigated using a probe station connected to a picoamperimeter model HP 4140B under ambient conditions. Figure 5.3 shows photographs of the probe station and of the resistors fabricated.

a   b

Figure 5.3. Photographs: (a) probe station and (b) electrical characterization of the SiC thin-film resistors.

The piezoresistive characterization of the resistors was carried out by beam-bending method using a simple experimental setup. In this method, a cantilever beam with one end clamped and another end free is used. It is known that when a load is applied to the free end of the beam the maximum stress occurs at the clamped end (see Figure 5.4). In order to maximize the piezoresistive effect, the resistor should be placed in this region of maximum longitudinal stress. In the experiments carried out in this work, it was used stainless steel cantilever beams with the following dimensions: L=120 mm, w= 25 mm and t= 1.2 mm.

One SiC resistor was glued with epoxy to 5 mm of the clamped end of the cantilever beam as shown in Figure 5.5 and cured at temperature of 120 °C for 30 min. Calibrated weights (20, 40, 60, 80 and 100 g) were applied to the free end of the beam (see Figure 5.6). A digital multimeter was used to measure the electrical resistance of the resistor without applied load on the beam and during subsequent tensile loading.

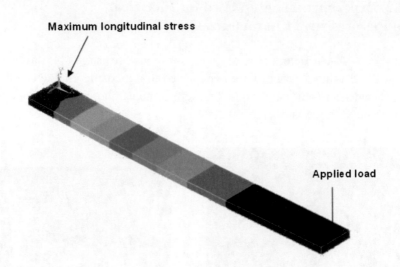

Figure 5.4. Distribution of longitudinal stress on a cantilever beam with one end clamped and another end free.

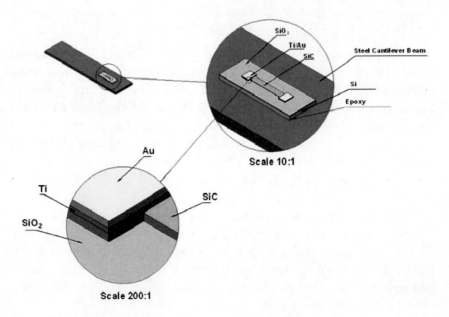

Figure 5.5. Schematic illustration of the SiC thin-film resistor glued on the beam.

The fractional electrical resistance change is determined by:

$$\frac{\Delta R}{R} = \frac{R_f - R_0}{R_0} \tag{5.2}$$

Where R_0 and R_f are is the electrical resistances without and with applied load on the beam, respectively. Three measurements were performed at each resistor type for each applied load.

The piezoresistive coefficient of the SiC thin-film resistor was obtained from equation that describes the piezoresistive effect [79]:

$$\frac{\Delta R}{R} = \pi_l (1 - \upsilon)\sigma_l \tag{5.3}$$

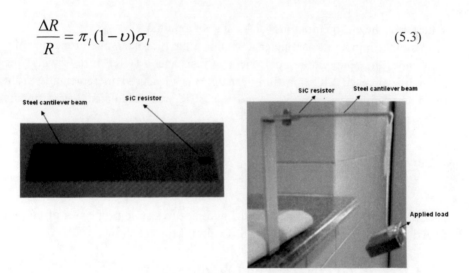

Figure 5.6. Photographs of the experimental set-up used to determine the GF of the resistors fabricated.

Where π_l is the longitudinal piezoresistive coefficient, υ is the Poisson's coefficient and σ_l is the longitudinal mechanical stress. In the case of the cantilever beam clamped at one end and free at the other, the longitudinal mechanical stress is defined by [80]

$$\sigma_l = \frac{6FL}{bt^2} \tag{5.4}$$

Where F is the weight of the block placed on free end of the beam, L, b and t are the length, width and thickness of the beam, respectively.

The resistance change can also be expressed in terms of strain using the gauge factor that is a dimensionless quantity and is given by:

$$GF = \frac{\Delta R}{R} \frac{1}{\varepsilon} \qquad (5.5)$$

Where ε is the strain that can be calculated by:

$$\varepsilon = \frac{\sigma_l}{E} \qquad (5.6)$$

Where E is the elastic modulus (or Young's modulus).

Another important parameter to evaluate the piezoresistive properties of a material is the temperature coefficient of resistance (TCR). In this work, the resistance of each SiC thin-film resistor was measured incrementally from room temperature up to 250°C and the TCR was calculated based on the following equation:

$$TCR = \frac{\Delta R}{R} \frac{1}{\Delta T} \qquad (5.7)$$

In the next sections, the electrical and piezoresistive properties of the un-doped and nitrogen-doped PECVD Si_xC_y thin films are shown.

5.2.3. Results on Electrical and Piezoresistive Characterization of Test Structures

The resistors fabricated on each type of film are numbered in the sequence shown in the Figure 5.7 (a). In this figure is also shown a photograph of the resistors fabricated on un-doped PECVD SiC film. To distinguish between these resistors, the letter N was added to the numbering of resistors fabricated on the nitrogen-doped PECVD film.

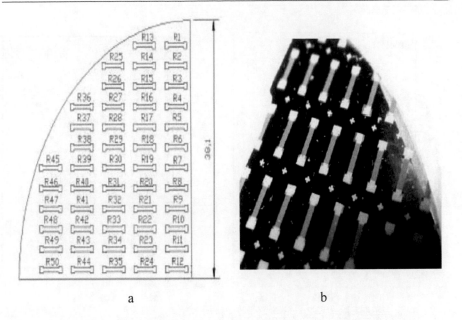

a   b

Figure 5.7. PECVD SiC thin-film resistors: (a) numeration sequence and (b) photograph.

In the I-V measurements of the un-doped resistors (sample P4) was used an initial voltage of 5V and final of 25V with step of 0.5 V while that for the nitrogen-doped resistors (sample P5) was used an input range of 0 to 1.8 V with step of 0.1 V.

In Figures 5.8 and 5.9 are shown I–V characteristics of the resistors fabricated. As can be observed un-doped resistors have a non-linear behavior. On the other hand, linear characteristics were observed in the nitrogen-doped films, which show that these resistors exhibit ohmic behavior. This indicates the nitrogen doping of PECVD SiC films during deposition process changed their electrical characteristics.

After the electrical measurements, it was performed the piezoresistive characterization of the thin-film resistors fabricated using the experimental set-up shown in section 5.2.2.

Figure 5.10 shows the change in resistance of un-doped and nitrogen-doped PECVD SiC thin-film resistors as a function of applied stress. The resistance constantly increases as the mechanical stress increases for both resistors.

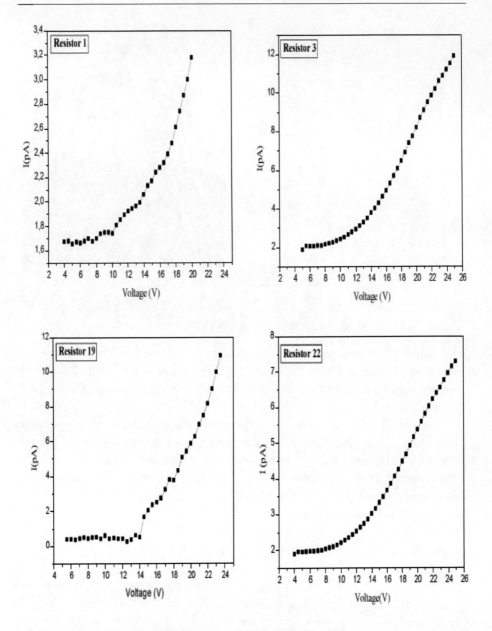

Figure 5.8. I-V characteristics of the PECVD un-doped SiC thin-film resistor.

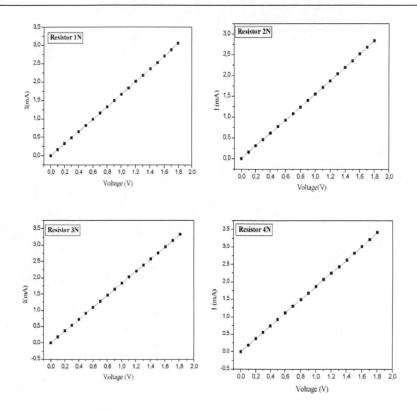

Figure 5.9. I-V characteristics of the PECVD doped SiC thin-film resistor.

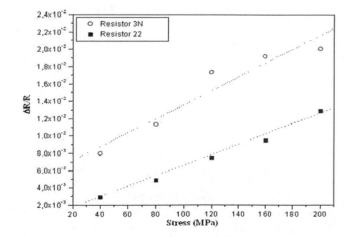

Figure 5.10. The fractional resistance change of the resistors 22 (un-doped PECVD SiC) and 3N (nitrogen-doped PECVD SiC) as a function of the applied stress.

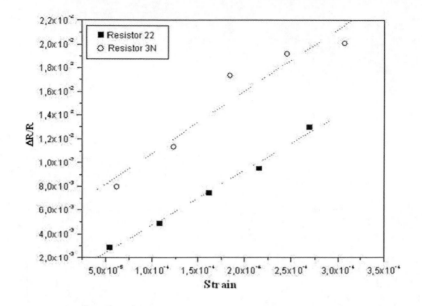

Figure 5.11. Gauge factor of the resistors 22 (un-doped PECVD SiC) and 3N (nitrogen-doped PECVD SiC).

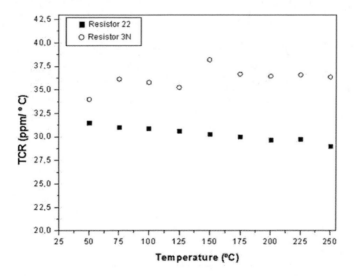

Figure 5.12. TCR measurements of the resistors 22 (based on un-doped PECVD SiC) and 3N (based on nitrogen-doped PECVD SiC).

Figure 5.11 shows the gauge factor (GF) obtained by plotting the relative change in resistance ($\Delta R/R$) as a function of the applied strain the thin-film resistors. It was observed that the GF of un-doped PECVD film is 49 whereas the doped is 47 [81].

The TCR measurements are presented in Figure 5.12. It can be observed that both resistors exhibited a positive TCR with values smaller than the based on silicon, which show their potential for high temperature applications.

5.3. DEVELOPMENT OF A PROTOTYPE OF PIEZORESISTIVE PRESSURE SENSOR BASED ON a-SiC FILM

5.3.1. Design of the Prototype

The design of a piezoresistive pressure sensor begins with the choice of the diaphragm geometry. Among different shapes of diaphragms, the square and circular are the most used. There are advantages and disadvantages of opting for one of these geometries. When comparing the response to pressure applied of a square diaphragm with a circular of similar dimensions, i.e. the side of the square diaphragm equals to the diameter of the circular, the stress is greater in the square diaphragm. However, in the circular diaphragm the stress is more evenly distributed making it easier placement of piezoresistors [82].

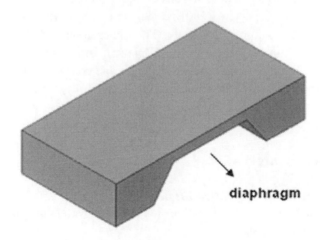

diaphragm

Figure 5.13. Cut view of the proposed sensor.

The sensor prototype presented here has a square diaphragm as shown in the schematic illustration in Figure 5.13. After defining the diaphragm geometry, the following hypotheses were considered for the physical model of the sensor:

- The behavior of the diaphragm of the sensor follows the theory of elasticity [83] and the diaphragm support structure has an infinite stiffness;
- The design of the diaphragm shall conform to the *thin plate or small deflection* theory [84], i.e. its maximum deflection is less than 1 / 5 of the thickness of the diaphragm;
- Poisson's ratio remains constant through the thickness of the diaphragm;
- There is not deformation in the medium plane of the diaphragm.

For the prototype developed in this book was chosen a layout, with six piezoresistors configured in Wheatstone bridge on a square diaphragm, following the model proposed by L. F. Fuller and S. Sudirgo [85] in order to place the piezoresistors where the stress on the diaphragm is the highest when pressure is applied. This optimized design allows maximizing the sensitivity of the sensor.

As the diaphragm of sensor will be fabricated by silicon etching in KOH solution, the mask-defined backside opening is larger than the actual diaphragm size. Given that KOH etches along (111) plane; thus, it makes 54.73° angle with respect to (100) Si substrate. The backside opening was designed to be 2500 μm by 2500 μm. The dimensions of the diaphragm are 1800 μm x 1800 μm x 30 μm. Two SiC piezoresistors have L= 700 μm and W= 100 μm and the others piezoresistors each one consists of two resistors in series with L= 350 μm and W= 100 μm (see Figure 5.14). The sensor chip size is 4.5 mm x 4.5mm.

5.3.2. Experimental

5.3.2.1. Fabrication of the Prototype

a) 2 inch (100) p-type Si double side polished wafers were RCA cleaned and thermally oxidized;
b) Photolithography on the backside of the wafer to pattern the diaphragm;

c) Anisotropic etching of Si in KOH solution;
d) Deposition of the SiC film on front side of the wafer;
e) Photolithography on the front side of the wafer to pattern the SiC piezoresistors;
f) The SiC film on the regions do not protected with photoresist was etched by reactive ion etching using SF_6+O_2 gas mixtures;
g) Photolithography on the front side of the wafer to pattern the Ti/Au electrical contacts;
h) Deposition of Ti and Au and lift-off.

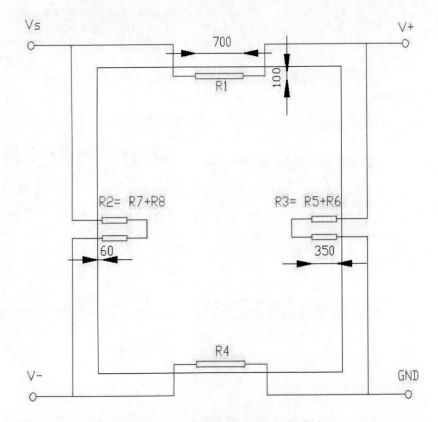

Figure 5.14. Configuration and dimensions of the piezoresistors on the diaphragm.

Mariana Amorim Fraga

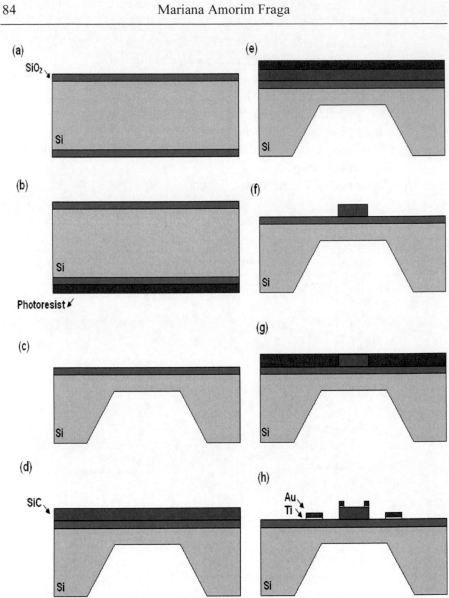

Figure 5.15. Illustration of the fabrication sequence of the piezoresistive pressure sensor prototype.

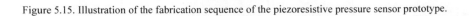

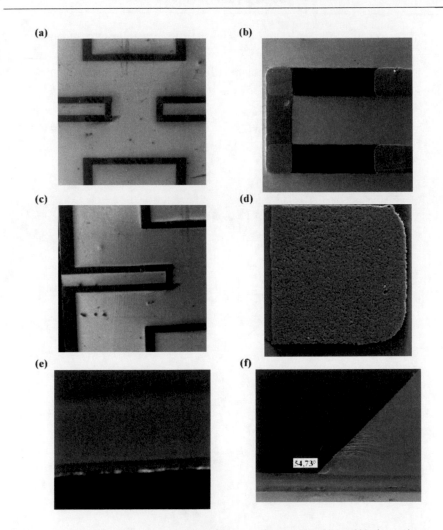

Figure 5.16. SEM images of the sensor: (a) six SiC piezoresistors (50X), (b) two SiC piezoresistors with Ti/Au metal lines (200X), (c) two SiC piezoresistors with Ti/Au metal lines (200X), (d) Ti/Au electrical contact (500X), (e) and (f) cross section of the SiO_2/Si diaphragm.

5.3.2.2. Package of the Prototype

The main functions of packaging are to protect the sensor from environmental factors such as humidity, light, and vibration besides allows the transmission of signals and provides mechanical support.

The sensor prototype presented in this book was packaged using a simple and cheap solution. The sensor die was glued on an alumina substrate (with Al metal lines) using silicone (Dow Corning 3140). An aluminum cup was used to protect the sensor.

The development of this package (illustrated in Figure 5.17) consisted of three steps: (1) Die attach, (2) Wire bonding, and (3) Glue Al cup.

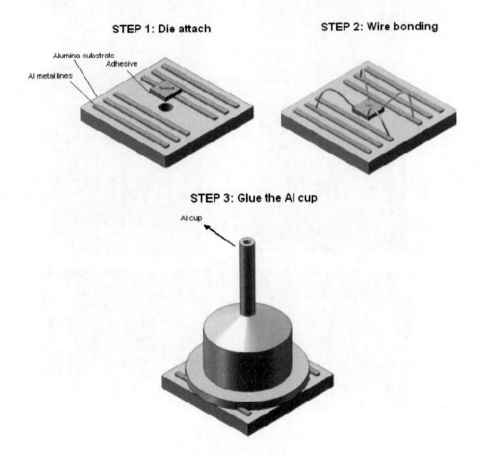

Figure 5.17. Steps of packaging of the sensor prototype.

The choice of the aluminum cup to package is due to easy machining of this material and reliability since it provides hermetic sealing of the device. This means that the device is totally isolated from environmental interference. The dimensions of Al cup used are shown in Figure 5.18. The photograph of the packaged sensor is shown in Figure 5.19.

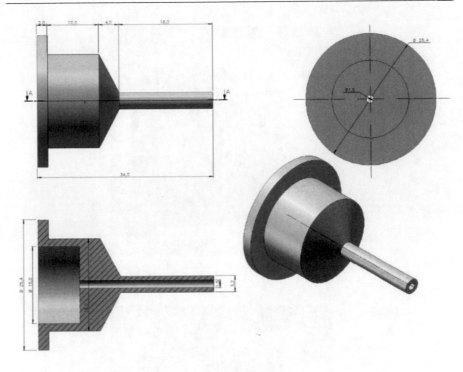

Figure 5.18. Dimensions of the Al cup used to protect the sensor.

5.3.2.3. Characterization of the Prototype

The output voltage of the sensor prototype was measured for applied pressure ranging from 0 to 12 psi and voltage suply of 12V. In Figure 5.20 is shown the output voltage as a function applied pressure on the sensor at room temperature. It was obtained an offset of 62 mV and a sensitivity of 0.33 mV/V.psi.

Table 5.1 shows the sensitivity and offset of three sensors fabricated on nitrogen-doped PECVD Si_xC_y film. For each sensor were performed, three measurements were taken at intervals of 10 days between the 1st and 2nd measurement and 45 days between the 2nd and 3rd.

a

b

Figure 5.19. Photographs of the sensors: (a) sensor chip glued to alumina substrate with Al metallization and (b) packaged sensor.

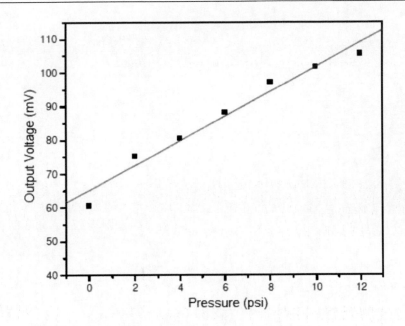

Figure 5.20. Output voltage of the sensor as a function of applied pressure.

Table 5.1. Characterization of the sensors

	Sensor 1			Sensor 2			Sensor 3		
Measure-ment	1[a]	2[a]	3[a]	1[a]	2[a]	3[a]	1[a]	2[a]	3[a]
Sensivity (mV/V. psi)	0.316	0.358	0.267	0.299	0.292	0.298	0.411	0.416	0.328
Offset (mV)	60.7	64.6	59.2	54.3	53.8	54.1	45.8	41.4	47.6

Chapter 6

FINAL REMARKS

This book is the result of 6 years of research on amorphous silicon carbide (a-SiC) thin films. In this period, all steps of synthesis, characterization, and processing of these films were studied.

The a-SiC films were produced by PECVD (plasma enhanced chemical vapor deposition) e RF magnetron sputtering techniques. The films obtained by two techniques were submitted to thermal annealing under argon atmosphere at 1000°C for 1h.

The as-deposited and annealed samples were investigated by Rutherford backscattering spectrometry (RBS), Raman spectroscopy, Fourier transform infrared spectroscopy (FTIR), x-ray diffraction (XRD), atomic force microscopy (AFM), four points probe, nanoindentation, and transmission/reflection measurements.

It was observed that the a-SiC films obtained by PECVD SiC are non-stoichiometric compounds of Si and C with high concentrations of carbon from 82% to 68%. FTIR and Raman spectra of these films indicated that the thermal annealing process changed the structure of chemical bonds. However, XRD spectra showed that only the sample deposited under lower SiH_4/CH_4 flow ratio presents SiC characteristic peaks visible which can be indexed as the (1 1 1) and (2 0 0) reflections of polycrystalline β-SiC. It was also observed that the resistivity did not change significantly with the reduction of carbon in the SiC film and the thermal annealing, but it was reduced by two orders of magnitude by the addition of nitrogen during the deposition process (in situ doping). The addition of nitrogen also reduced the elastic modulus of the PECVD SiC films. Furthermore, the elastic modulus decreased with the carbon concentration in the films. On the other hand, it was observed that the film composition did not change the optical band-gap.

With respect to SiC films produced by RF magnetron sputtering, it was observed that the sample deposited in an argon environment is a stoichiometric compound of Si and C and that the introduction of nitrogen gas to the deposition process promoted the formation of the ternary compound silicon carbonitride (SiC_xN_y). It was observed that as nitrogen flow increases there is a significant reduction in the concentration of carbon in the film. In addition, the nitrogen incorporation in the film was over 45% comproving that in the deposition conditions used the nitrogen did not act as dopant and promoted the formation SiC_xN_y. The resistivity and optical gap of the films also indicated the formation of this compound since these parameters were not reduced by the introduction of nitrogen.

A comparison between the properties of the a-SiC films produced show that the obtained by PECVD have most attractive characteristics for development of piezoresistive sensors. Given this, the piezoresistive characterization was performed only with the samples P4 (un-doped) and P5 (nitrogen-doped). The GF and TCR measured confirm the potential of a-SiC films as an alternative material to silicon for high temperature piezoresistive sensors applications.

The prototype of piezoresistive pressure sensor based on a-SiC film developed has a good sensitivity similar to those showed by sensors based on crystalline or bulk SiC.

It is noted that the methodology presented in this book can be applied to the development of piezoresistive sensors based on others types of semiconductor thin film materials.

REFERENCES

[1] R.S. Okoije, A. Ned, A.D Kurtz, "α(6H)-SiC Pressure Sensors for High Temperature Applications", *IEEE sensors,* pp.146-149, (1996).

[2] M. Willander, M. Friesel, Q. Wahab,B. Straumal, "Silicon carbide and diamond for high temperature device applications", *Journal of Materials Science: Materials in Electronics,* v.17, pp.1-25, (2006)

[3] W. Shockley, Proceedings of the First International Conference on Silicon Carbide, Boston, MA, (1959).

[4] M. Madou, "Fundamentals of Microfabrication". CRC press, New York, (1997).

[5] G. Kroetz., E. Obermeier, W. Wondrak, C.Cavalloni, "Silicon Carbide on Silicon – An Ideal Material Combination for Harsh Environment Sensor Applications", *IEEE MEMS,* pp.732-736, (1998).

[6] R. Singh; L.L. Ngo; S.H. Seng.; F.N.C. Mok ; "A Silicon Piezoresistive Pressure Sensor", *Proceedings of the IEEE,* (2001).

[7] J.S. Shor, D. Goldstein, A.D. Kurtz, "Characterization of n-Type β-SiC as a Piezoresistor", *IEEE Transactions on Electron Devices,* pp.1093-1099, (1993).

[8] A.F. Flannery, J.N. Mourlas, C.W. Storment, S. Tsai, S.H. Tan, G.T.A.Kovacs , "PECVD Silicon Carbide for Micromachined Transducers", *IEEE Transducers,* pp.217-220, (1997).

[9] H. Wu, H. Stefanescu, C. A.Zorman, M. Mehregany M., "Fabrication and Testing of Single Crystalline 3C-SiC piezoresistive Pressure Sensors", Eurosensors XV, (2001).

[10] N. A. E. Forhan, M. C. A. Fantini, I. Pereyra, "Nano-Crystalline Si1-xCx:H thin films deposited by PECVD for SiC-On-Insulator application", *Journal of Non-Crystalline Solids,* pp.338-340, (2004).

[11] S. Limpijumong, "Theoretical Study of Some Aspects of Polytypism in Silicon Carbide", Reports of the Departament of Physics Case Western Reserve University (01/2000).

[12] Y. Baba, T. Sekiguchi, I. Shimoyama, K. G. Nath, "Structures of sub-monolayered silicon carbide films", *Applied Surface Science* 237, pp. 176-180, (2004).

[13] P. G. Neudek, "SiC Technology", NASA report, pp. 1-54 (1998).

[14] J.W. Palmour, A. Agarwal; S. H. Ryu, M. Das, J. Sumakeris and A. Powell, "Large Area Silicon Carbide Power Devices on 3 inch wafers and Beyond", Cree Inc. report, (2004).

[15] J. Richmond, S. Hooge and J. W. Palmour, "Silicon Carbide Power Applications and Device Roadmap", Cree Inc report., (2004).

[16] E. K. Polychroniadis, A. Andreadou, A. Mantzari, "Some recent progress in 3C-SiC growth. A TEM Characterization" *Journal of Optoelectronics and Advanced Materials* Vol. 6, No. 1, pp. 47 – 52 (2004)

[17] R. W. Barry, P. M. Hall, M. T. Harris, J. Klerer, "Thin Film Technology", Journal of the Electrochemical Society, V.117, (1970)

[18] R. R. Rapozo, "Estudo da deposição de filmes finos de carbeto de silício em substrato de grafite", Monografia de conclusão do curso de Engenharia Mecânica Aeronáutica, Instituto Tecnológico de Aeronáutica (2004).

[19] O. A Weinreich and A. Ribner, "Optical and Electrical Properties of SiC Films Prepared in a Microwave Discharge", *Journal of Electrochemical Society,* V.115, 1090, (1968).

[20] T. E. Hartman, J. C. Blair, C. A. Mead, "Electrical conduction through thin amorphous SiC films", *Thin Solid Films,* V.79, pp. 79-93, (1968).

[21] C. J Mogab and W.E. Kingery, "Preparation" *Journal Applied Physics,* 39, 3640 (1968).

[22] H. Matsunami," Progress in epitaxial growth of SiC", *Physica B,* vol. 185, pp. 65, (1993)

[23] M. A. Fraga, H. Furlan, M. Massi, I. C. Oliveira "Effect of nitrogen doping on piezoresistive properties of a-SixCy thin film strain gauges" *Microsystem Technologies,* Vol.16, pp.925-930, (2010).

[24] M. A. Fraga, H. Furlan, M. Massi, I. C. Oliveira, L. L. Koberstein, "Fabrication and characterization of a SiC/SiO2/Si piezoresistive pressure sensor", *Procedia Engineering.* Vol. 5, pp.609-612, (2010).

[25] M. Ohring, The Materials Science of Thin Films (Academic Press Inc. 1992)

[26] I. Doi, "Técnicas de deposição: CVD, notas de aula da disciplina IE726 – Processos de Filmes Finos", Universidade Estadual de Campinas, (2006).

[27] S. Wolf and R. N. Tauber, "Silicon Processing for the VLSI", V. 1, California, Lattice Press , (2000).

[28] J. J. Hanak and J. P. Pellicane, "Effect of secondary eletrons and negative ions on sputtering films", *Journal Vac. Sci. Technol.* , vol. 13, p.406, (1976).

[29] S. Pascoli, "Obtenção e caracterização de filmes de TiO2 depositados sobre cerâmica de revestimento via magnetron sputtering DC", Tese de Doutorado apresentada a Universidade Federal de Santa Catarina, (2007).

[30] B. Chapman, "Glow Discharges Processes", John wileys & Sons , New York, (1976)

[31] W. E. Spear and P. G. Le Comber, "Substitutional doping of amorphous silicon", *Solid State Communications,* V. 17, pp.1193-1196, (1975)

[32] D. A. Anderson and W. E. Spear, "Electrical and optical properties of amorphous silicon carbide, silicon nitride and germanium carbide prepared by the glow discharge technique", *Philosophical. Magazine,* V. 36, pp. 1-16, (1977).

[33] Y.Tawada, K. Tsuge, M. Kondo, H. Okamoto, and Y. Hamakawa, "Properties and structure of a-SiC:H for high-efficiency a-Si solar cell", *Journal Applied Physics,* 53, (1982).

[34] F. Demichelis, G. Crovini, C. F.Pirri, E. Tresso, R. Galloni, C. Summonte, R. Rizzoli, F. Zignani and P. Rava, "Boron and phosphorus doping of a-SiC:H thin films by means of ion implantation", *Thin Solid Films,* V. 265, pp. 113-118, (1995).

[35] W. Chu, J. W. Mayer, M. Nicolet, "Backscattering Spectrometry", Academic Press, New York, (1978).

[36] L. R. Doolittle, "A semiautomatic algorithm for rutherford backscattering analysis" , Nuclear Instruments and Methods in Physics Research Section B, V.15, pp.227-231, (1986).

[37] J. Huran, B. Zaťko, P. Boháček, A. P. Kobzev, A. Vincze, Ľ. Malinovský and A.Valovič "Properties of hydrogenated amorphous/nanocrystalline carbon films prepared by plasma enhanced chemical vapour deposition", *Innovations in Thin Film Processing and Characterisation,* pp.1-4, (2009)

[38] B. C. Smith, "Fundamentals of Fourier transform infrared spectroscopy", CRC Press, (1996).

[39] N. B. H. D. Colthup and S. E. Wilberley, "Introduction to Infrared and Raman Spectroscopy", New York: Academic Press, (1964).

[40] J. R. Ferraro and K. Nakamoto, "Introductory Raman Spectroscopy", New York: Academic Press, (1984).

[41] E. W. Nuffield, " X-ray diffraction methods", John Wiley, New York, (1966).

[42] H. P. Klug and L. E. Alexander, "X-ray diffraction procedures", John Wiley & Sons, New York (1974).

[43] G.Binnig, C. F. Quate, C. Gerber, "Atomic Force Microscopy", Physical Review Letters, V. 56, (1986).

[44] W. R. Runyan, "Semiconductor Measurements and Instrumentation", McGraw-Hill, New York, (1975).

[45] A. P. Schuetze, W. Lewis, C. Brown, W. J. Geeters, "A laboratory on the four-point probe technique", *American Journal Physics,* V. 72, (2004).

[46] W. C. Oliver and G. M. Pharr, "A new improved technique for determining hardness and elastic modulus using load and sensing indentation experiments", *Journal of Materials Research,* V.7, pp.1564-1582, (1992).

[47] A. R. Oliveira, "Dopagem elétrica de filmes finos de carbeto de silício amorfo hidrogenado (a-SiC:H) obtidos por PECVD", Dissertação de Mestrado apresentada a Escola Politécnica da USP, (2002).

[48] S. T. Pantelides, "Theory and modeling of SiC oxidation", Department of Physics and Astronomy, Vanderbilt University, (2000)

[49] L. Porter and R. Davis, "A critical review of ohmic and rectifying contacts for silicon carbide", *Materials Science Engineering B,* vol. 34, pp. 83-105, (1995).

[50] S. M. Sze, "VLSI Technology", McGraw-Hill, (1988).

[51] J. L. Weyner, S. Lazar, J. Borysiuk, J. Pernot, "Defect-selective etching of SiC", *Physica Status Solidi,* V. 202, Issue 4, pp. 578 – 583, (2004).

[52] L. Jiang, R. Cheung, R. Brown, A. Mount, "Inductively coupled plasma etching of SiC in SF6/O2 and etch-induced surface chemical bonding modifications", *Journal of Applied Physics,* Vol. 93, No. 3, pp. 1376-1383, (2003).

[53] P. H. Yih and A. J. Steckl, "Effects of Hydrogen Additive on Obtaining Residue-Free Reactive Ion Etching of β-SiC in Fluorinated Plasmas", *Journal Electrochemical. Society,* V. 140, pp.1813-1824, (1993).

[54] J. P. Li, P. H. Yih and A. J. Steckl, "Thickness Determination of SiC-on-Si Thin Films by Anisotropic Reactive Ion Etching and Preferential Wet Etching", *Journal Electrochemical. Society,* V.140, pp.178-182, (1993).

[55] M. A. Fraga, R. S. Pessoa, M. Massi, H. S. Maciel, S. G. Santos Filho, Etching Studies of Post-Annealed SiC Films Deposited by PECVD: *Influence of the Oxygen Concentration ECS Transactions,* V. 9, pp.227-231, (2007).

[56] I. Behrens, E. Peiner, A. S. Bakin, A. Schlachetzki, "Micromachining of silicon carbide on silicon fabricated by low-pressure chemical vapour deposition", *Journal Micromechanics and Microengineering,* V.12, pp. 380-384 (2002).

[57] H. Furlan, "Desenvolvimento de Membranas para Sensores de Pressão Utilizando Freamento Eletroquímico", Tese de Doutorado apresentada a Escola Politécnica da USP, (2003).

[58] I. Pereyra, M. N. P. Carreno, "Wide gap a Si1-xCx:H thin films obtained under starving plasma deposition conditions", *Journal of Non-Crystalline Solids,* V. 201, pp.110-118, (1996).

[59] A. Hammad, E. Amanatides, D. Mataras, D. Rapakoulias, "PECVD of hydrogenated silicon thin films from SiH4+H2+Si2H6 mixtures*", Thin Solid Films,* V. 451-452, pp. 255-258, (2004).

[60] Y. P. Guo, J. C. Zheng, A. T. S. Wee, C. H. A. Huan, K. Li, J. S. Pan, Z. C. Feng, S. J. Chua, *"Photoluminescence studies of SiC nanocrystals embedded in a SiO2 matrix"* , V. 339, pp. 319-322, (2001).

[61] T. Young, C. Hsiao, C. Peng, "Characterization of SiC thin films grown on Si by inductively coupled plasma chemical vapor deposition at low temperatures", *Semiconducting and Insulating Materials Conference,* SIMC-XI, (2000).

[62] M. A. Fraga, R. S. Pessoa, M. Massi., H.S. Maciel, S.G. Santos Filho, "Synthesis and Etching of amorphous SixCy thin films with high carbon content*, Brazilian Journal of Vaccum Applications,* V. 26, pp.193-197, (2007).

[63] M. A. Fraga, R. S. Pessoa, M. Massi, H.S. Maciel, S.G. Santos Filho, "Etching Studies of Post-Annealed SiC Film Deposited by PECVD: Influence of the Oxigen Concentration*", ECS Transactions,* Vol. 9 (1), pp. 227-234, (2007).

[64] C. J. Mogab, A.C. Adams, and D.L. Flamm, "Plasma ctching of Si and SiO2 — The effect of oxygen additions to CF4 plasmas*", Journal of Applied Physics,* Vol. 49, pp.3796, (1978).

[65] K. B. Sundaram and J. Alizadeh, " Deposition and optical studies of silicon carbide nitride thin films", *Thin Solid Films,* Vol. 370, 1-2, pp.151-154, (2000).

[66] J. Alizadeh and K. B. Sundaram,"Electrical studies on amorphous silicon carbide nitride films", *Journal of Materials Science Letters,* Vol. 21, pp. 7-8, (2002).

[67] S. Komatsu, et al., "Preparation and Characterization of Reactively Sputtered SiCxNy Films", *Thin Solid Films,* Vol. 193/194, pp. 917-923, (1990).

[68] K. B. Sundaram, J. Alizadeh, R. M. Todi and V. H. Desai, " Investigations on hardness of rf sputter deposited SiCN thin films", *Materials Science and Engineering A,* Vol. 368, pp.103-108, (2004).

[69] P. Chabert, G. Cunge, J.-P. Booth and J. Perrin, "Reactive ion etching of silicon carbide in SF6 gas: Detection of CF, CF2, and SiF2 etch products", *Applied Physics Letters,* Vol. 79, pp. 916-918, (2001).

[70] M. A. Fraga, R. S. Pessoa, M. Massi, I. C. Oliveira, H.S. Maciel, S.G. Santos Filho, "Etching Characteristics and Surface Morphology of nitrogen-doped a-SiC films prepared by RF magnetron sputtering", *ECS Transactions,* V.14, pp. 375-382, (2008).

[71] S. Sugiyama, M.Takigawa, and I. Igarashi, "Integrated Piezoresistive Pressure Sensor with Both Voltage and Frequency Output", *Sensors and Actuators A,* Vol. 4, pp. 113–120, (1983).

[72] A.J. Shaw, B. A. Davis, M. J. Collins, L.G. Carney, "A Technique to Measure Eyelid Pressure Using Piezoresistive Sensors", *IEEE Transactions on Biomedical Engineering,* pp. 2512-2517, (2009).

[73] A.L. Window, G.S. Holister, Strain Gauge Technology (Elsevier Applied Science, London New York, 1982).

[74] J. M. Stauffer, B. Dutoit, B. Arbab, " Standard MEMS Sensor Technologies for Harsh Environment", *IEEE IET seminar on MEMS sensors and actuators,* Vol. 1, pp. 91–96, (2006).

[75] R.S. Okojie, A.A. Ned, A.D. Kurtz, "Operation of (6H)-SiC pressure sensor at 500°C", *International Conference on Solid State Sensors and Actuators, Vol. 2,* pp. 1407-1409, (1997).

[76] R. Ziermann, J. von Berg, W.Reichert, E. Obermeier, M. Eickhoff, G. Krotz, "A high temperature pressure sensor with β-SiC piezoresistors on SOI substrates *", International Conference on Solid State Sensors and Actuators, Vol. 2,* pp. 1411-1414, (1997).

[77] C.H. Wu , S. Stefanescu, H. I. Kuo, C. A. Zorman , M. Mehregany, "Fabrication and testing of single crystalline 3C-SiC piezoresistive

pressure sensors" *Proceedings of the 11th International Conference on Solid-State Sensors and Actuators,* (2001).

[78] M. A. Fraga, M. Massi , H. Furlan , I. C. Oliveira, L. A. Rasia , C. F. R. Mateus, "Preliminary evaluation of the influence of the temperature on the performance of a piezoresistive pressure sensor based on a-SiC film", *Microsystem Technologies,* in press, (2011).

[79] R. Singh, L. L. Ngo, S.H. Seng and F. N. Mok , "A Silicon Piezoresistive Pressure Sensor". Proceedings of the IEEE, (2001).

[80] F. P. Beer and E. R.Johnston, Mechanics of materials, 2nd edn. McGraw-Hill, London, (1992).

[81] M. A. Fraga, H. Furlan, M. Massi, I. C. Oliveira,, "Effect of nitrogen doping on piezoresistive properties of a-SixCy thin film strain gauges", *Microsystem Technologies* 16, 925-930, (2010)

[82] M. A. Fraga, "Projeto de um Sensor de Pressão Piezoresistivo em substrato SOI", *Dissertação de Mestrado apresentada a Escola Politécnica da USP* (2005).

[83] J. W. Dally, Experimental Stress Analysis. McGraw-Hill Book Company, (1978).

[84] S. Timoshenko and S. Woinoswky-Krieger, Theory of plates and Shells. McGraw- Hill, (2001)

[85] L. F.Fuller and S. Sudirgo, "Bulk micromachined pressure sensor", Proceedings of IEEE 15th Biennial University/Government/Industry Microelectronics Symposium, (2003).

INDEX

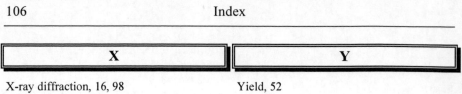